Defect structure in nanomaterials

Related titles:

Corrosion protection and control using nanomaterials
(ISBN 978-1-84569-949-9)
The book is divided into two parts. Part one looks at the fundamentals of corrosion behaviour and the manufacture of nanocrystalline materials. Chapters discuss the impact of nanotechnology in reducing corrosion cost, and investigate the influence of various factors including thermodynamics, kinetics and grain size on the corrosion behaviour of nanocrystalline materials. There are also chapters on electrodeposition and the corrosion behaviour of electrodeposited nanocrystalline materials. Part two provides a series of case studies of applications of nanomaterials in corrosion control. Chapters review oxidation protection using nanocrystalline structures at various temperatures, sol- gel and self-healing nanocoatings and the use of nanoreservoirs and polymer nanocomposites in corrosion control.

Nanostructured metals and alloys
(ISBN 978-1-84569-670-2)
Nanostructured metals and alloys reviews the latest technologies used for the production of nanostructured metallic materials, as well as recent advances in research into their structure and mechanical properties. Part one describes the different methods used to process bulk nanostructured metals and alloys, including chapters on severe plastic deformation, mechanical alloying and electrodeposition among others. Part two concentrates on the microstructure and properties of nanostructured metals, with chapters studying deformation structures such as twins, microstructure of ferrous alloys by equal channel angular processing, and characteristic structures of nanostructured metals prepared by plastic deformation. In part three, the mechanical properties of nanostructured metals and alloys are discussed, with chapters on such topics as strengthening mechanisms, nanostructured metals based on molecular dynamics computer simulations, and surface deformation. Part four focuses on existing and developing applications of nanostructured metals and alloys, covering topics such as nanostructured steel for automotives, steel sheet and nanostructured coatings by spraying.

Anti-abrasive nanocoatings
(ISBN 978-0-85709-211-3)
This book provides an overview of the fabrication methods for anti-abrasive nanocoatings. The connections among fabrication parameters, nanocoatings characteristics and the resulting properties (i.e. nanohardness, toughness, wear rate, load-bearing ability, friction coefficient, and scratch resistance) are discussed. Size-affected mechanical properties of nanocoatings are examined, including their uses. Anti-abrasive nanocoatings including metallic-, ceramic-, and polymeric-based layers as well as different kinds of these nanostructures, such as multi-layered nanocomposites and thin films, are reviewed.

Defect structure in nanomaterials

Jenő Gubicza

WOODHEAD PUBLISHING

Oxford Cambridge Philadelphia New Delhi

Published by Woodhead Publishing Limited, 80 High Street, Sawston,
Cambridge CB22 3HJ, UK
www.woodheadpublishing.com
www.woodheadpublishingonline.com

Woodhead Publishing, 1518 Walnut Street, Suite 1100, Philadelphia, PA 19102–3406, USA

Woodhead Publishing India Private Limited, G-2, Vardaan House, 7/28 Ansari Road,
Daryaganj, New Delhi – 110002, India
www.woodheadpublishingindia.com

First published 2012, Woodhead Publishing Limited

British Library Cataloguing in Publication Data
A catalogue record for this book is available from the British Library.

Library of Congress Control Number 2012938315

Woodhead Publishing ISBN 978-0-08-101660-2 (print)
ISBN 978-0-85709-614-2 (online)

Typeset by RefineCatch Limited, Bungay, Suffolk

Contents

List of figures

List of tables

Preface

Nanostructured materials (or nanomaterials) are built up from microstructural units (e.g. grains) having smaller size than 100 nm at least in one dimension of space. This class of materials includes thin layers that have thicknesses smaller than 100 nm, while their lateral dimensions are usually several centimetres, which means that the criterion for nanostructured materials is fulfilled only in one dimension. Nanotubes also pertain to nanomaterials as their diameter is several nanometres, i.e. they satisfy the criterion in two dimensions. Crystalline nanoparticles and bulk materials, having grain sizes smaller than 100 nm, are also members of this class, although these materials are also called as nanocrystalline materials since their microstructural unit (the crystalline grain or particle) is smaller than 100 nm in all directions. It is noted that in some cases the size of microstructural unit depends on the method used for studying the microstructure. For instance, in severely deformed metallic materials the grain size measured by electron microscopy is usually several hundreds of nanometres while the crystallite size obtained from the broadening of x-ray diffraction peak profiles is smaller than 100 nm. This apparent dichotomy is attributed to the fact that in these materials the crystallite size corresponds rather to the subgrain size. Therefore, these specimens are often classified as ultrafine-grained materials for which the average grain size is between 100 and 1,000 nm but the term of nanomaterials is also used for these samples as the subgrain size is smaller than 100 nm. In the last decade of the second millennium, nanomaterials have become a focal point of materials science due to their unique physical, chemical and mechanical properties that destine these materials to novel and promising applications. The small dimension of the grains or particles in nanomaterials and their specific processing methods affect their defect structure (vacancies, dislocations, planar defects and grain boundaries), which has a significant influence on the properties of these materials. The knowledge of the relationships between the production methods, the lattice defects and the physical properties of nanomaterials is very important not only in order to understand the specific phenomena occurring when the grain size is very small, but also from the point of view of practical applications of these materials.

This book aims to synthesise the knowledge of lattice defects formed in nanomaterials either in their production or during subsequent straining and storing of the as-received specimens. First, the processing methods of nanomaterials are overviewed and then the lattice defect structures formed during the synthesis of nanomaterials are characterised. Special attention is paid to the lattice defects in low stacking-fault energy nanomaterials. The thermal stability of defect structure in nanomaterials is also studied. The influence of lattice defects on mechanical and hydrogen storage properties is discussed in detail. X-ray line profile analysis is an effective tool for studying the defect structure in nanomaterials, therefore a short description of this method is provided in the appendix. The results are organised and presented in a form that is hopefully beneficial for a wide audience: materials scientists and engineers as well as lecturers and students at universities.

About the author

Jenő Gubicza was born in 1969 in Budapest, Hungary. He studied physics and mathematics at Eötvös Lorand University (Budapest, Hungary) and graduated in 1992. He received his PhD in physics at the same university in 1997 with the qualification *summa cum laude*. He worked in the Research Institute for Technical Physics and Materials Science of the Hungarian Academy of Sciences from 1996 to 1998. He returned to Eötvös Lorand University in 1998. He got his habilitation degree in 2005. Dr Gubicza earned the scientific title of Doctor of the Hungarian Academy of Sciences in 2009. Now he is working in the Department of Materials Physics at Eötvös Lorand University as an associate professor.

Dr Gubicza's main research field is the study of the defect structure in nanomaterials by x-ray diffraction line profile analysis. He has contributed to the development of a new evaluation method of diffraction peak profiles. Dr Gubicza used this method for investigating the relationship between the processing conditions, the defect structure and the mechanical properties of nanocrystalline and ultrafine-grained materials. He revealed the effect of the melting point and the stacking-fault energy on the evolution of the dislocation density and the related yield strength in ultrafine-grained fcc metals processed by severe plastic deformation. Dr Gubicza also showed the correlation between the sintering conditions (e.g. time, temperature, pressure and atmosphere) and the defect structure in nanostructured metallic and ceramic materials. He extensively investigated the thermal stability of nanomaterials and revealed the influence of the low stacking-fault energy on the decrease of the stability of ultrafine-grained microstructures processed by severe plastic deformation.

Dr Gubicza has published two book chapters and more than 140 journal papers that have been cited more than 1,100 times. He was a member of the Physics Jury of the Hungarian National Scientific Fund (2005–08) and the Council of Faculty of Science at Eötvös Lorand University (2005–06). Presently Dr Gubicza is the Chairman of the Diffraction Section of Roland Eotvos Physical Society, a member of the Solid State Physics Committee of the Hungarian Academy of Sciences and the Scientific Committee of the Hungarian Conference Series on Materials Science. He was awarded the Schmid Rezso Prize of Roland Eotvos Physical Society in 2008 and the Bolyai-plaquette of the Hungarian Academy of Sciences in 2010. He was a member of the organising committees of several international conferences and the originator and co-organiser of the unique international PhD courses on x-ray line profile analysis held in 2009 and 2011.

Dr Gubicza teaches materials science at Eötvös Lorand University but he was also invited to give lectures in Budapest University of Technology and Economics, Hungary. His lectures also cover the basic theory and the different practical applications of x-ray diffraction. Dr Gubicza has supervised more than 10 MSc and PhD theses. He has written textbooks in materials science and x-ray diffraction at Eötvös Lorand University and the University of Miskolc, Hungary. He was invited as a visiting professor to Université Paris 13 in 2006, 2009, 2010 and 2011.

1

Processing methods for nanomaterials

Abstract: In this chapter, the processing methods of bulk nanomaterials are briefly overviewed. The techniques presented here are used for production of the materials studied in the later chapters of this book. These processing methods include 'top-down' procedures that produce nanomaterials by severe plastic deformation as well as 'bottom-up' approaches where the nanograins are built up from individual atoms or from their clusters.

Key words: severe plastic deformation, powder metallurgy, electrodeposition, crystallisation of bulk amorphous materials.

1.1 Processing of bulk nanomaterials by severe plastic deformation

The processing methods of bulk nanomaterials are usually classified into two groups, as depicted in Figure 1.1. In the course of 'bottom-up' methods the materials are built up from individual atoms, molecules or their clusters (particles), such as in electrodeposition or inert gas condensation. In the case of 'top-down' methods, the nanosized microstructural units (grains or crystallites) are achieved by refinement of

Figure 1.1 Classification of processing routes of bulk nanomaterials

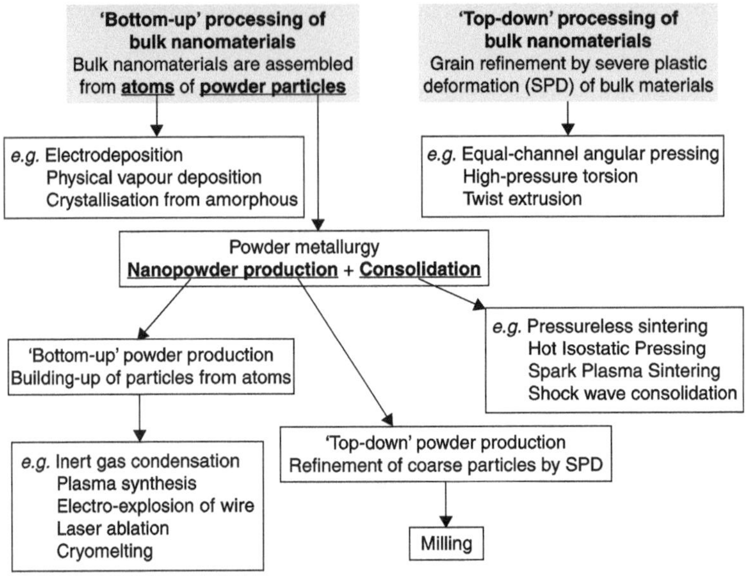

coarse-grained materials. The grain refinement usually occurs by severe plastic deformation that can be carried out either on bulk materials, as in the case of equal-channel angular pressing, or on powder samples, e.g. by milling. In the case of nanomaterials processed by powder metallurgy, the classification may be more complex. The consolidation procedures of nanopowders are usually declared as 'bottom-up' methods, but the nanopowders used for sintering can be produced by milling that is a 'top-down' procedure.

Bulk ultrafine-grained (UFG) or nanomaterials can be produced by severe plastic deformation (SPD) of their coarse-grained counterparts. Formally, processing by severe plastic deformation (SPD) may be defined as those metal forming procedures in which a very high strain is imposed on a bulk solid without the introduction of any significant change in the overall dimensions of the solid [1]. As a result of SPD, dislocations are formed and arranged into low-energy configurations (e.g. low-angle grain boundaries) that transform into high-angle grain boundaries with increasing strain [1, 2]. SPD-processing can be performed on both bulk materials and powder samples, as can be seen in the classification scheme of the processing methods of nanomaterials in Figure 1.1. In this section, an overview of the SPD methods applied on bulk materials is given. A solid is considered a bulk if its size is in the millimetre range or larger in any direction. The most important advantage of bulk SPD techniques compared with other processing methods (e.g. powder metallurgy) of nanomaterials is that they avoid undesired contamination and porosity in the final microstructure. The initial material is usually a bulk coarse-grained workpiece that is deformed plastically at an equivalent strain higher than 1. The equivalent strain in a deformation corresponds to the strain in a uniaxial tension that yields the same plastic work as produced in the given deformation.

The most frequently applied SPD procedure is *equal-channel angular pressing* (ECAP) [1, 3–5], also known as equal-channel angular extrusion (ECAE). The principle of ECAP is illustrated in Figure 1.2a [5]. A circular rod or a square bar is pressed through a die containing two intersecting channels. The cross-sections of the channels and the billet match, therefore deformation takes place only at the intersection of the two lubricated channels. Since the cross-sectional dimensions of the billet remain unchanged, the pressings may be repeated to attain exceptionally high strains. It is noted that after removal of the billet from the exit channel, its diameter increases slightly due to an elastic relaxation. Therefore, if the entry and the exit channels have the same cross-sections, a surface layer of the billet should be removed between the consecutive passes. However, this

Figure 1.2 (a) The schematic depiction of ECAP-processing; and (b) the four fundamental processing routes in ECAP.

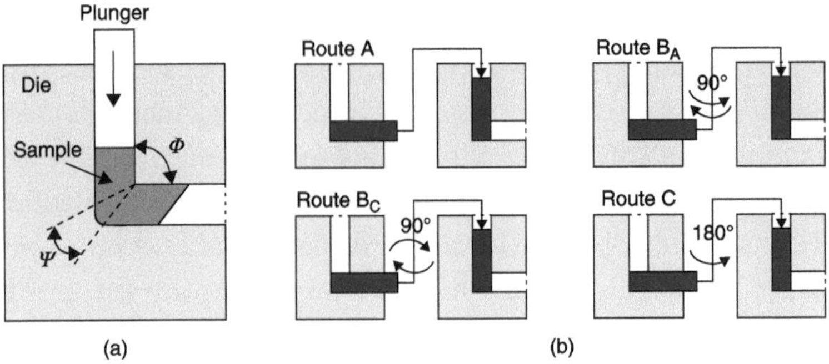

Source: Reprinted from Nakashima et al. (2000) [7] with permission from Elsevier

problem may be overcome if the exit channel exhibits a slightly smaller diameter (by about 0.1 mm) than for the entry channel. In the latter case, the billet is extruded when it passes through the intersection of the two channels. The equivalent strain, ε, introduced in ECAP is determined by a relationship incorporating the angle between the two parts of the channel, Φ, and the angle representing the outer arc of curvature where the two parts of the channel intersect, Ψ. The relationship is given by the formula [1]:

$$\varepsilon = N\frac{2ctg\left(\frac{\Phi}{2}+\frac{\Psi}{2}\right)+\Psi\sin^{-1}\left(\frac{\Phi}{2}+\frac{\Psi}{2}\right)}{\sqrt{3}}, \qquad (1.1)$$

where N is the number of passes. In practice, the channel angle Φ ranges from 45° to 180° and the arc of curvature varies from 0° to 90° [1]. The usual values of Φ and Ψ are 90° and 20°, respectively, and in this configuration one pass corresponds to an equivalent strain of ~1 [6]. The imposed strain increases proportionally with increasing number of passes. Between the consecutive passes, the billet can be rotated about its longitudinal axis. According to the various manners of rotation, four different basic routes of ECAP can be distinguished, as depicted in Figure 1.2b. In the case of route A, there is no rotation of the billet, while in routes B_A or B_C rotations by 90° in alternate directions or the same direction are applied, respectively. In route C the billet is rotated by 180° about the longitudinal axis. The number and orientations of shear planes are different for various routes [1, 7], which affects the evolution of the microstructure during ECAP. When using a die with a channel angle of $\Phi = 90°$, route B_C is generally the most expeditious way to develop a UFG microstructure consisting of homogeneous and equiaxed grains with grain boundaries having high angles of misorientation [1]. The advantages of ECAP are that the as-processed workpiece (i) has relatively large dimensions (several centimetres) in all directions and after several passes the UFG microstructure exhibits (ii) a high degree of homogeneity and (iii) a large fraction of equilibrium high-angle grain boundaries. The latter feature gives improved fatigue behaviour and a good corrosion resistance [8].

Another frequently studied SPD method is *high-pressure torsion* (HPT). High-pressure torsion refers to processing in which the sample, generally in the form of a thin disk, is subjected to torsional straining under a high hydrostatic pressure: the principle of HPT is illustrated schematically in Figure 1.3 [2, 5]. The disk is located within a cavity, a hydrostatic pressure is applied and plastic torsional straining is performed by rotation of one of the anvils. The torsion of the disk is achieved due to the friction between the sample and the anvils. The unique feature of this method is the extremely large applied pressure, 1–9 GPa [2], which hinders the annihilation processing of lattice defects, thereby yielding to very large defect densities and small grain size. Usually,

Figure 1.3 The principle of HPT

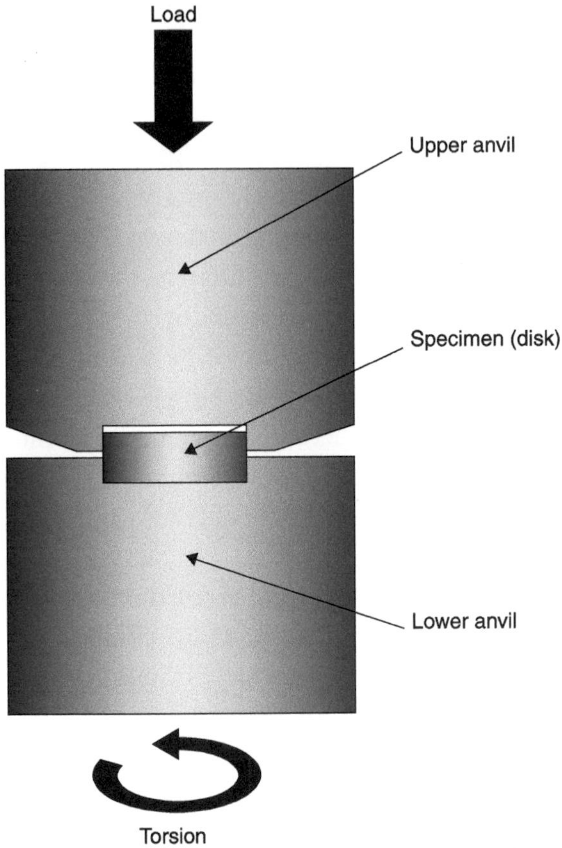

the thickness and the diameter of the HPT-processed disks are not larger than about 1 and 10 mm, respectively, which limits their practical applications. The equivalent strain (ε) depends on the distance from the centre (r) according to the equation:

$$\varepsilon = \frac{2\pi r N}{\sqrt{3}h}, \tag{1.2}$$

where h is the height of the disk and N is the number of revolutions [2]. It is noted that usually there is some outward flow of material between the two anvils during HPT and the value of *h* is reduced accordingly, which should be taken into account in the calculation of the strain.

The SPD-processing of circular rods or rectangular bars by *multi-directional forging* (MDF) includes multiple repeats of forging operations with changes of the axes of the applied load [9, 10]. The principle of MDF is illustrated in Figure 1.4 for the case of a rectangular workpiece. After one cycle of MDF, the original shape of the billet is regained and its

Figure 1.4 Schematic illustration of the steps of MDF procedure. The axes of the reference system attached to the sample are denoted as *x*, *y* and *z*. The loading directions are indicated by black arrows. The letters *a*, *b* and *c* denote the sample dimensions

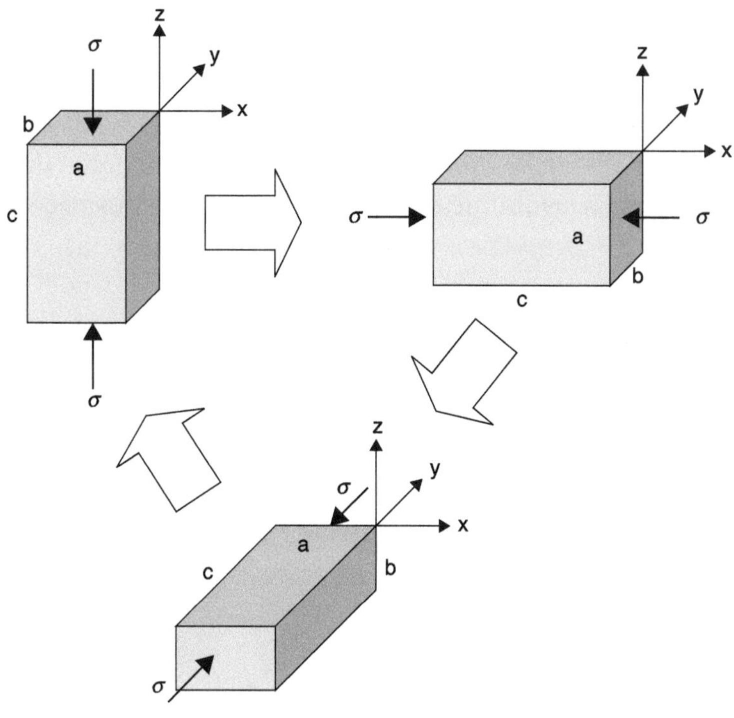

longitudinal axis has the same orientation in the reference system attached to the sample as before deformation. In the case of a circular rod, several forging steps (about 20) are usually needed to restore the rod shape of the sample after one cycle. As in one forging step, the imposed strain is about 0.2, therefore one cycle corresponds to an equivalent strain of ~4. The choice of the appropriate temperature-strain rate regimes of MDF leads to the desired grain refinement. The operation is usually realised over the temperature interval of 0.1–0.5 T_m, where T_m is the absolute melting temperature. MDF is useful for producing large-sized billets with nanocrystalline or UFG microstructures [5].

During *twist extrusion* (TE), a workpiece is pushed through a twisted die that yields torsion of the sample with a designated angle around its longitudinal axis (see Figure 1.5). The workpiece regains its shape and size after each TE pass, therefore it is possible to repetitively process a sample for stronger grain refinement [5, 11]. By analogy to HPT, the plastic strain is not uniform across the cross-section but rather increases with the distance from the twist axis, therefore the more distant regions have a finer grain size. This microstructural heterogeneity leads to inhomogeneous mechanical properties: the hardness increases as the distance from the centre in the cross-section increases. The homogeneity of the microstructure can be improved by increasing the number of TE passes.

Figure 1.5 The principle of TE processing

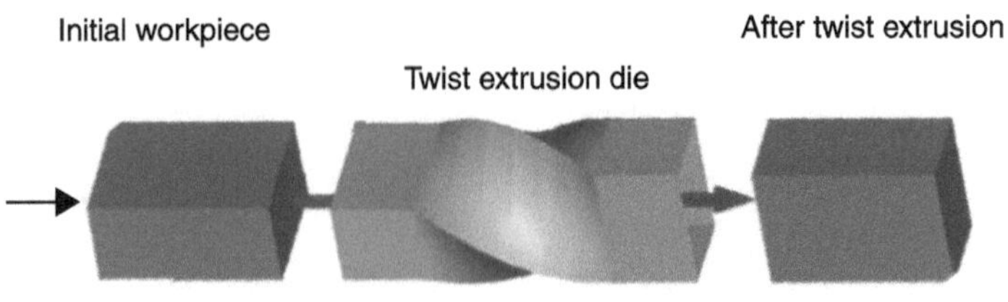

Figure 1.6 The principle of the ARB method

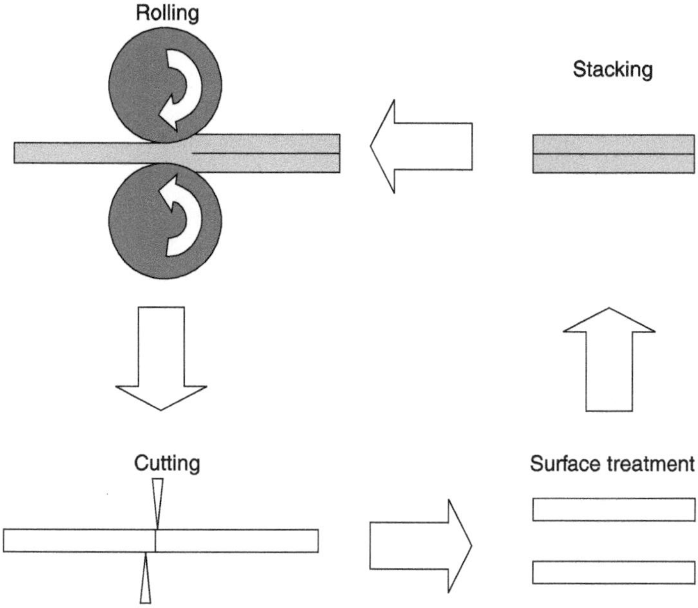

During *accumulative roll-bonding* (ARB), first a sheet is rolled so that the thickness is reduced to one-half of the initial value (see Figure 1.6). The rolled sheet is cut into two halves that are stacked together. The stacked sheets are then rolled again to one-half thickness. The repetition of rolling, cutting and stacking operations yields to a large accumulated strain in the sheet. To achieve good bonding during the rolling step, the surfaces of the sheets are degreased and wire-brushed before placing them in contact [5]. In practice, the UFG grains produced by ARB are not equiaxed but rather they have a pancake-like shape which is elongated in the rolling direction. The most important advantage of this method is its high productivity and that it can be performed by a conventional rolling facility. It is noted that beside ECAP, HPT, MDF, TE and ARB methods, numerous other SPD procedures exist for processing bulk UFG or nanocrystalline materials.

Figure 1.7 The principle of the RCS method

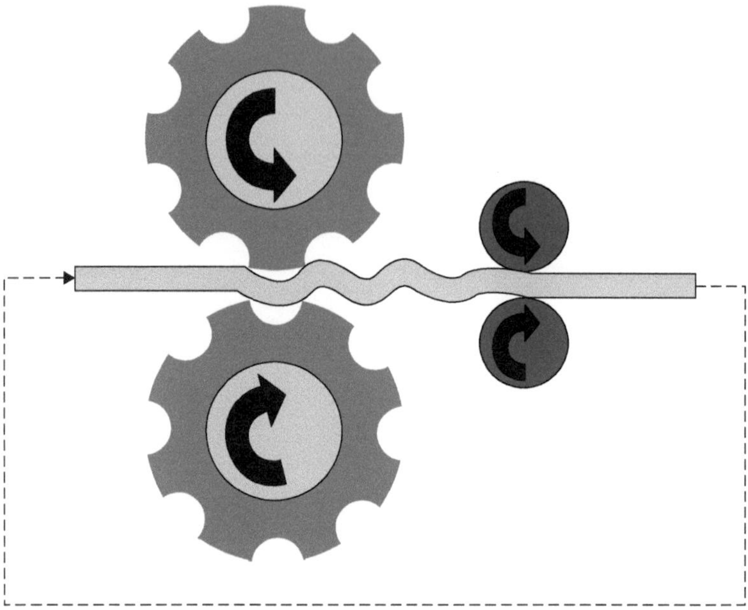

In the process of *repetitive corrugation and straightening* (RCS), the workpiece is initially deformed to a corrugated shape and then straightened, which may be repeated many times, as depicted in Figure 1.7. An advantage of RCS is that it can be adapted easily to current industrial rolling facilities and therefore it has the potential of producing nanostructured materials in a continuous and economical way [5].

1.2 Processing of nanomaterials by powder metallurgy

The processing of bulk nanomaterials by powder metallurgy includes the following basic steps: (i) nanopowder production, (ii) compaction of the powder into a high porosity specimen having the same shape as the final product (forming step) and (iii) consolidation of the sample to high density usually at high pressure and temperature (sintering step). The second step is especially important in the case of brittle materials, such as ceramics, since after sintering the manufacturing of the very hard and rigid workpiece to the desired form is difficult without fracture. Although during the forming step the sample receives the final shape, its size is still larger than that of the final product due to the large porosity. This specimen is called a 'green body'. In the sintering step, the shape of the sample does not change, only its size is reduced as a consequence of the decrease of porosity. For tough metallic materials, the forming step is often missing, since the workpiece can be manufactured even after sintering.

The driving force of nanopowder consolidation during sintering is the reduction of the large surface area of nanoparticles. Additionally, the high sintering pressure also assists the consolidation by inducing large shear stresses at the contact points between particles that yield plastic deformation, thereby contributing to the filling of pores. This mechanism is especially important in the case of metallic powders having good deformability. Although the high temperature during sintering facilitates the consolidation by increasing the atomic mobility, it may also cause grain coarsening, which is an unwanted phenomenon in processing of nanomaterials. In order to minimise grain growth, usually the time and temperature of sintering are reduced together with a simultaneous increase of pressure for maintaining the low porosity level in the final product. The most important advantages of powder metallurgy methods compared with SPD procedures are that (i) usually smaller grain size can be achieved, (ii) the thermal stability is better and (iii) the lack of texture gives an isotropic behaviour. At the same time, the drawbacks are (i) the small productivity, (ii) the remaining porosity and (iii) the undesired contamination. In the following, first some nanopowder production procedures are presented and then the consolidation methods are reviewed.

The processing methods of nanopowders can be classified as 'bottom-up' and 'top-down' procedures. In the first case, the powder particles are built up from individual atoms or atomic clusters. In the second case, coarse-grained initial powder particles are subjected to severe plastic deformation, usually with the application of milling. As a result, UFG or nanocrystalline microstructure is formed inside the metallic particles, while the particle size remains in the micrometre range [12]. In the case of brittle materials (e.g. ceramics), the fracture of particles into smaller parts also operates during milling.

In the following, five 'bottom-up' nanopowder processing methods are presented: inert-gas condensation, laser ablation, cryogenic melting, radiofrequency plasma synthesis and electro-explosion of wire. In the process of *inert-gas condensation*, first atoms are thermally evaporated from a metallic source inside an ultra-high vacuum chamber filled with inert gas, typically helium (see Figure 1.8) [13, 14]. The vaporised species then lose their energy via collisions with helium molecules. As collisions limit the mean free path, supersaturation can be achieved above the vapour source that yields the formation of atomic clusters [14, 15]. The clusters are transported to a liquid-nitrogen-filled cold finger by a convection flow and then they are condensed as nanoparticles on the surface of the finger. The particles are removed from the cold finger by means of a scraper assembly. They are collected and transported to an in-situ compaction device. Consolidation is performed first in the low-pressure compaction unit and then in the high-pressure compaction unit [13]. When atoms are ejected from the target surface by

Figure 1.8 **Schematic picture of an inert-gas condensation facility**

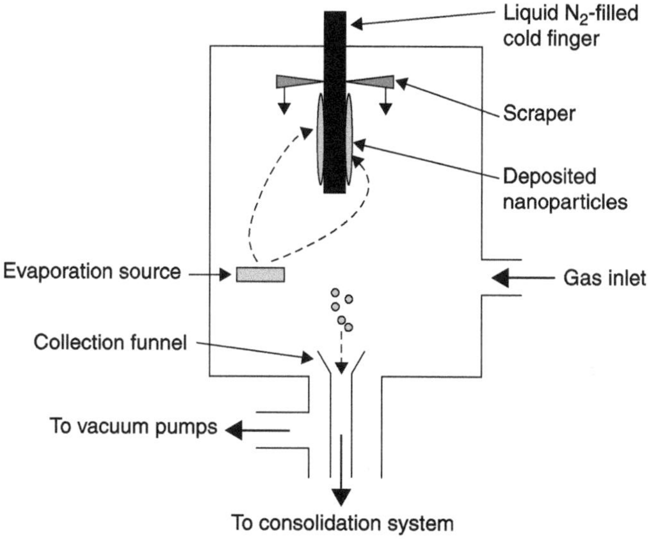

the impact of energetic ions, the process is called sputtering. Sputtering is capable of depositing high melting point materials such as Mo, Ta, W and ceramics, which are difficult to fabricate using evaporation. In the method of *laser ablation*, an intense pulsed laser beam irradiates the target of interest, thereby vaporising atoms and clusters from the target [13].

In the process of *cryogenic melting* (or *cryomelting*), a metal rod melts by a radio frequency inductor, as depicted in Figure 1.9 [13, 16]. The molten metal droplets fall into a cool gas flux evaporated from a cryogenic medium, such as liquid Ar at a temperature of 77 K. Rapid overheating of the metal via the radio frequency technique produces a substantial evaporation rate of atoms from the hot surface of the droplets into the cool gas. Then nanoparticles are formed by rapid condensation of the supersaturated metal vapour. The condensation region, where the particles are formed by

Figure 1.9 Schematic depiction of the apparatus for cryogenic melting

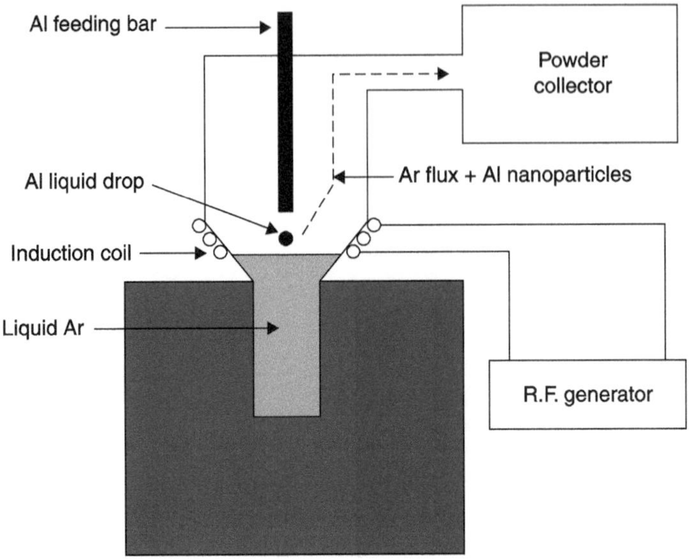

nucleation, growth and coalescence processes, is featured by a high temperature gradient, i.e. typically from 2,200 K at the metallic surface but drop to 77 K in the cryogenic medium. The low temperature of the surrounding medium produces a high rate of nucleation and a rapid cooling of the as-formed particles, which yields their small size. The upward cryogenic gas flux transports the nanoparticles into the powder collector.

The method of *electro-explosion of wire* is basically used to prepare metallic nanoparticles [17–19]. The facility for electro-explosion of wire process is illustrated in Figure 1.10. A very thin wire (about 0.5 mm) and a plate electrode are made of a metal and connected to a battery (tension ≈ 10 V). The electrical circuit remains open until the contact is made by the wire onto the plate. The very high current density (10^4–10^6 A/mm^2) in the thin wire results in an explosion of

Figure 1.10 Schematic diagram of the set-up for electro-explosion of wire: (1) thin metal wire electrode, (2) metal plate electrode, (3) batteries and (4) glass vessel

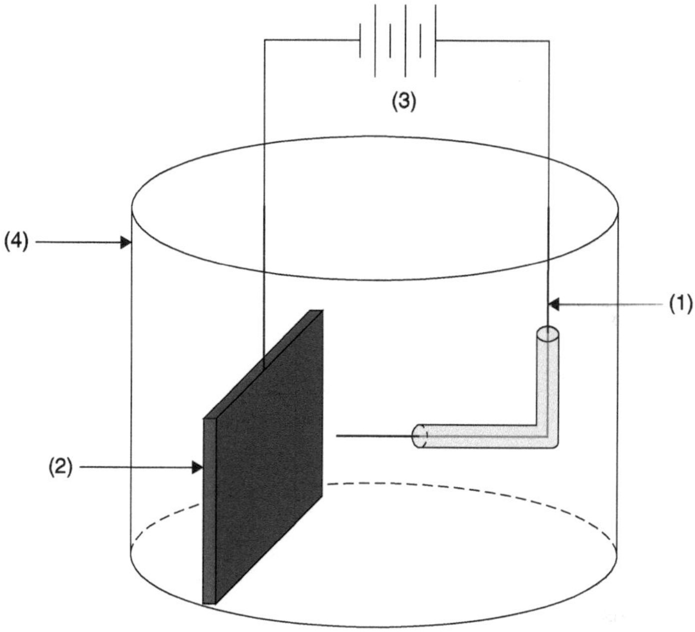

the wire in a very short time: the material boils up in a burst, a bright light flashes and a mixture of superheated vapour and boiling droplets scatter into the ambient atmosphere [17]. The scattered material is condensated into nanoparticles in the surrounding medium (e.g. in Ar gas). After explosion, the circuit opens again, and the feeding of the wire causes further explosion processes. The frequency of explosions is about 1 Hz.

Ball milling of powders is a 'top-down' procedure of nanopowder processing. The essence of the method is that coarse-grained powder together with balls made of hard materials (e.g. steel or ceramics) are filled into a mill. The collisions of the powder particles with the balls and the internal wall of the mill yield severe plastic deformation, which leads to grain-refinement inside the particles in a similar manner as in SPD-processing of bulk metallic materials. It must be noted that nanopowders do not necessarily consist of nanoparticles. The powder particles may actually be of micrometric sizes, yet the particle interior may be divided into many nanograins [12]. The advantages of ball milling techniques are the simplicity and the cost-effectiveness compared with other nanopowder production procedures. A disadvantage of the method is that the powder can be contaminated by the material of the balls, e.g. Fe contamination can occur due to steel balls. Additionally, the milling is often carried out in inert gas (e.g. in Ar) in order to minimise oxidation of particles.

There are many different designs of ball mills, which can be used for processing of nanopowders. Some often-applied equipment is: jar (drum) mill, Szegvari attritor, planetary mill, vibratory (shaker) mill and magnetic ball mill. The milling balls usually have a diameter between 5 and 10 mm, and the ball-to-powder weight ratio can vary from 1 to 100. In the case of the laboratory *jar mill* (or industrial drum mill), large numbers of grinding balls are placed inside a cylindrical container, which is rotated by rollers around a horizontal axis (see Figure 1.11). The balls may roll down the inside wall surface, which produces shear forces on powder trapped between the wall and the ball, but mostly they fall freely, accelerated only by gravitational force, then impacting the powders (and other balls) beneath them [12]. The jar mills are low-energy mills, although they can provide higher-energy milling if sufficiently large diameter in order of metres and many balls are used.

Larger acceleration than gravitational force, hence higher velocity and higher kinetic energy of balls, can be achieved in

Figure 1.11 Jar or drum mill

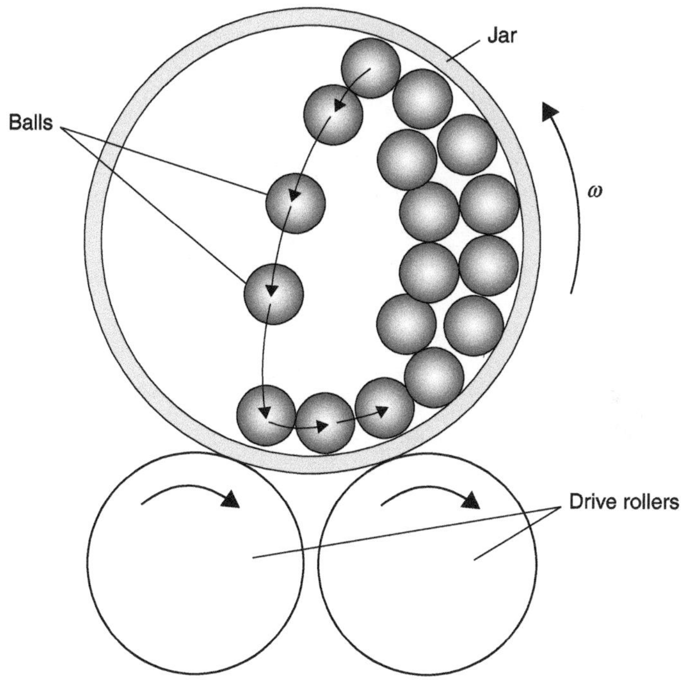

a *Szegvari attritor*. In such mills the milling is conducted in a cylinder filled with balls that are stirred by rotating impellers (see Figure 1.12). The impact of impellers causes differential velocities between the balls and the powder, therefore the powder particles are deformed mainly by shear. An attritor-type ball mill delivers 10 times more energy than a jar mill with comparable size, and attritors can be considered as medium-energy ball mills. The efficiency of vertical attritor mills is limited by the tendency of the powder to fall by gravity to the bottom of the cylinder [12].

In a *planetary ball mill* the jars containing the powder and the balls are rotating around their own axes with an angular velocity of ω_v, and simultaneously they orbit around a

Figure 1.12 Szegvari attritor

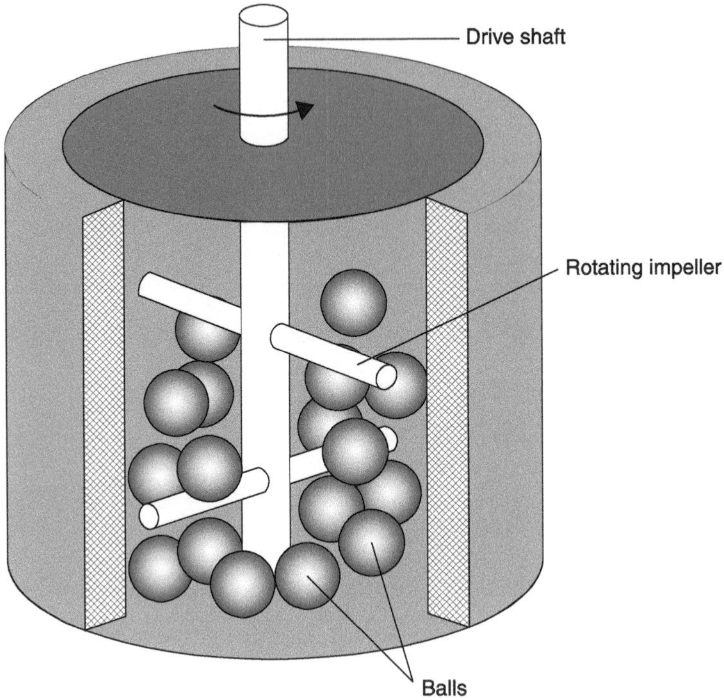

common axis with an angular velocity of ω_p, similarly as the movement of the planets around the Sun. The motion of the jars and the balls in a planetary mill is shown in the schematic in Figure 1.13. The impact forces achieved in planetary mills are usually higher than 50 times the gravitational force acting on the balls [12]. Planetary ball mills can be considered as medium- to high-energy ball mills; however, milling times needed to process submicron-sized and nanostructured powders may be long.

In *vibratory or shaker ball mills*, an eccentric motion is imparted to a cylindrical vial containing the powder and the balls at frequencies ranging from 10 to 2,000 Hz, and at small amplitudes of vibrations [12]. The balls oscillate in

Figure 1.13 Motion of balls and jars in a planetary mill

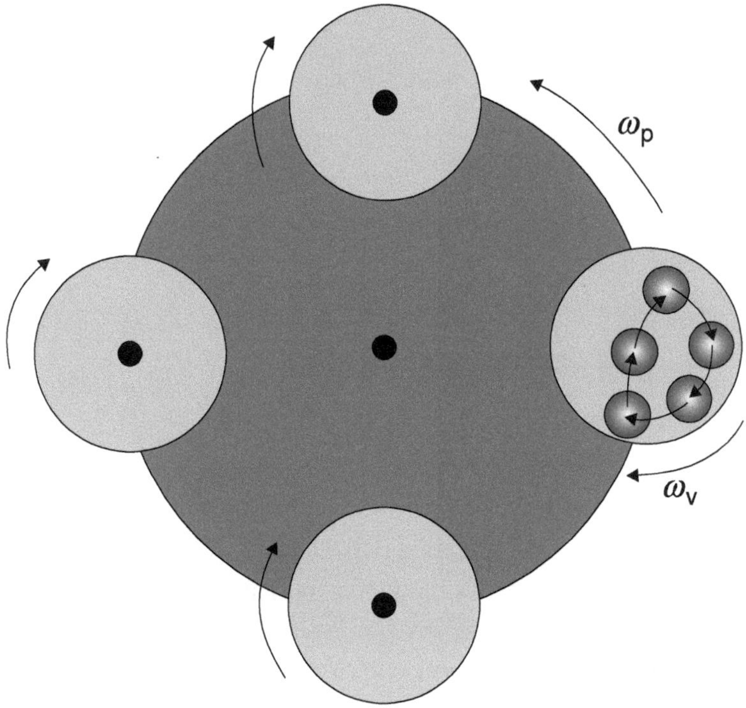

three, or more, mutually perpendicular planes within a small vial, as illustrated in Figure 1.14. Most of the energy transfer in vibratory mills is conducted in the mechanical impact mode, although substantial shear is also present on the powder particles trapped between the balls and the internal wall of the vial. This is a high-energy milling process, as the balls impact the powder at high speeds and at high frequencies. In order to prevent the powder against oxidation, the milling is usually performed in inert gas.

Another way to apply force to a milling ball, apart from the gravitational and the centrifugal means discussed earlier, is to drive ball motion by magnets. In the case of the *magnetic ball mill*, the magnetic balls are attracted to the inside surface

Figure 1.14 Illustration of a vibratory ball mill

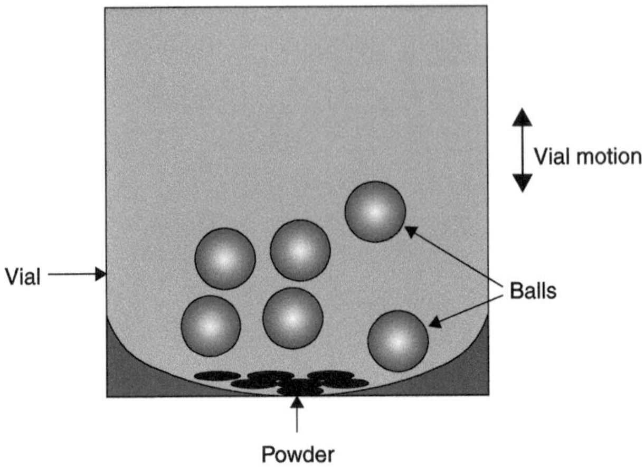

of the non-magnetic jar by strong permanent magnets placed outside the jar (see Figure 1.15). The pull imparted by the external magnet on the magnetic balls is so strong that the centrifugal force acting on the balls becomes a secondary factor in milling. The point where each ball is detached from the wall is well determined by the position of the magnet; hence each ball imparts the same energy impact on the milled powder. In the magnetic ball mill, high-energy milling can be achieved at low rotational speeds (30–200 rpm), therefore the Fe contamination can be reduced. The shear and impact forces can be controlled by the position of the external magnets [12].

In the following, some nanopowder consolidation procedures are overviewed: shock wave consolidation, pressureless sintering, hot pressing, hot isostatic pressing and spark plasma sintering. The reason for selecting these methods for review is that mainly they were used for the consolidation of the samples studied in the next chapters. In *shock wave consolidation*, the high pressure and rapid loading rates applied on the powder result in interparticle

Figure 1.15 Motion of balls in a magnetic mill

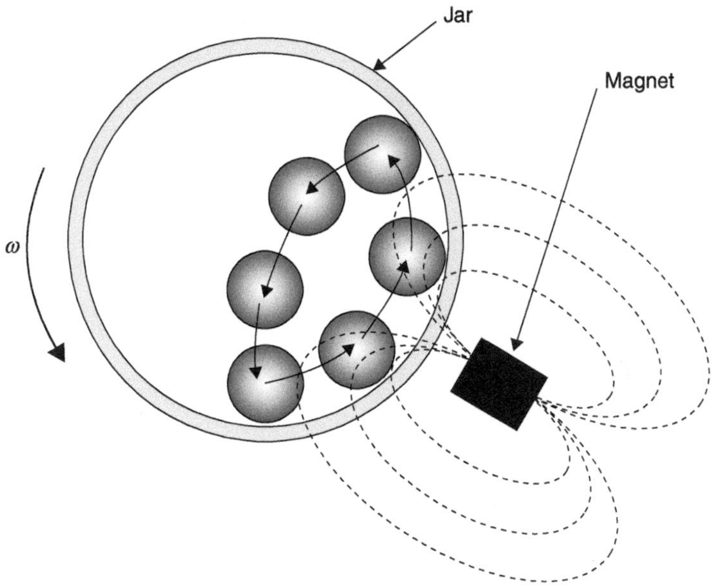

bonding due to localised melting at the interfaces between the particles [20, 21]. In practice, the powder particles are enclosed in a steel block which is covered by a plate on the top that drives the shock wave caused by the explosion towards the sample (see Figure 1.16). The driver plate is usually made of a highly conductive, highly ductile material, such as copper. Explosives and the detonator are packed on top of the driver plate. Of late, gas guns are also widely used in performing shock wave consolidation. In the latter case, hydrogen gas activates the projectile, which travels all the way through the barrel and hits the powder sample. The duration and the pressure of the shock wave, which yields consolidation of the powder, are ~10^{-6} s and 20–600 GPa, respectively. Shock wave consolidation technique is mainly used for metallic powders. Because only the surfaces of the particles are heated up, the grain growth is suppressed.

Figure 1.16 **Schematic depiction of shock wave consolidation process showing the experimental set-up**

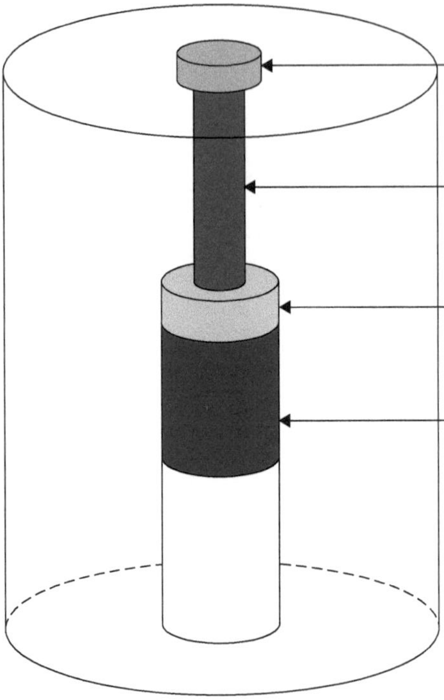

During *pressureless sintering*, the pre-compacted sample (green body) is kept at a high temperature for a long time, which yields fusion of the particles, thereby decreasing the porosity in the specimen. The driving force for consolidation is the high surface energy in the pre-compacted sample compared with the dense material. The surface area is reduced by diffusion on the particle surfaces, resulting in their fusion, as illustrated in Figure 1.17. Fusion occurs well below the melting point of the material, but at a temperature sufficiently high to allow an acceptable rate of diffusion to occur, usually at greater than one-half of the melting point on a Kelvin scale [20]. The high

Figure 1.17 **Illustration of the fusion of particles during pressureless sintering**

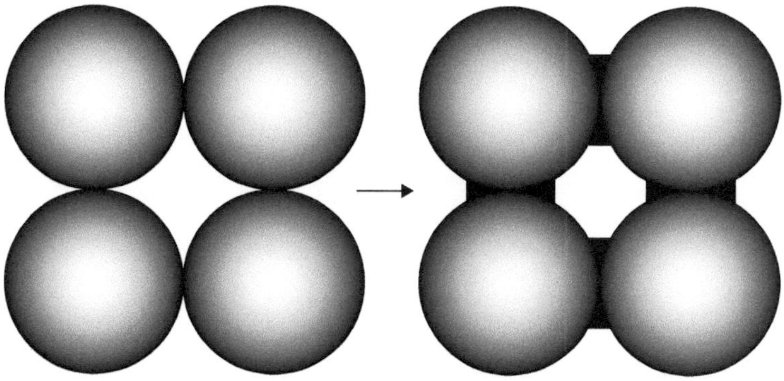

Source: Reprinted from Viswanathan et al. (2006) [20] with permission from Elsevier.

temperature and the relatively long time (hours) of pressureless sintering usually yields grain growth in the consolidated sample.

In *hot pressing* process, a uniaxial loading up to 25 tonne assists the consolidation at high temperature [20]. The schematic depiction of the equipment is presented in Figure 1.18. The application of high pressure enables the reduction of temperature and time of consolidation in order to decrease the grain growth compared with the pressureless sintering process. The external pressure induces internal stresses between the particles, which facilitates diffusion and also yields plastic deformation of the particles in the case of metallic powders, thereby speeding up the consolidation. A vacuum or inert atmosphere (e.g. Ar) can be applied to prevent oxidation.

In *hot isostatic pressing* (HIP), high hydrostatic pressure and high temperature are applied to consolidate fine particles [20, 22, 23]. If only high hydrostatic pressure is used and no heating is performed, the process is called *cold isostatic*

Figure 1.18 **Schematic depiction of hot pressing**

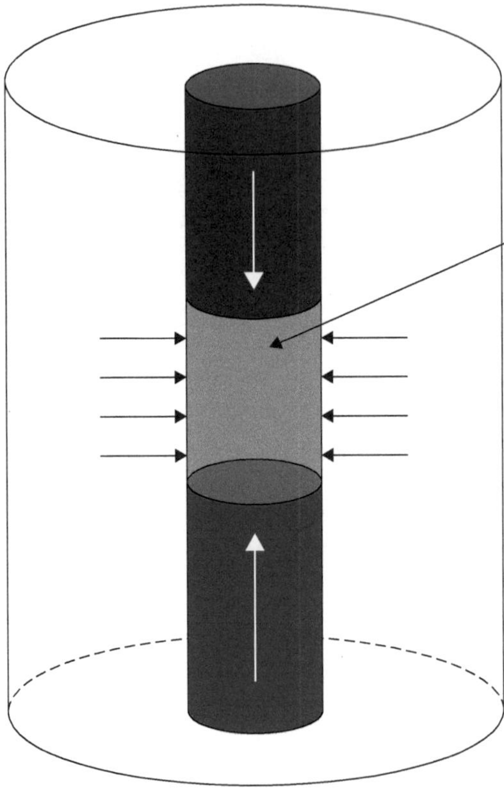

pressing. During the manufacturing process, the powder is placed in a container, typically a steel can. The container is subjected to a very high vacuum to remove air and moisture from the powder. This processing step usually takes a long time (up to 100 hours). The container is then sealed and subjected to HIP. The application of high inert gas pressures and elevated temperatures yields the removal of internal voids and creates a strong bond throughout the material. The result is a clean, homogeneous material with a uniformly fine grain size and a near 100% density. The reduced porosity of HIP-processed materials leads to improved mechanical properties and increased workability. One of the primary advantages of the HIP process is its ability to create near-net shapes that require little machining.

In *spark plasma sintering* (SPS), the consolidation is assisted by high current density pulses that enable the decreased time and temperature of sintering, thereby reducing grain growth. A schematic depiction of a typical SPS apparatus is shown in Figure 1.19, consisting of a graphite die where powder is loaded and is subjected to high electric current [20, 24, 25]. The high current density results in spark discharge (high temperature plasma) between the gaps of particles, which activates the surface by removing surface oxide. This leads to faster diffusion on the surfaces of the purified particles, resulting in easier consolidation. Additionally, the high current density at the contact points of the particles yields a local melting of the particles' surfaces,

Figure 1.19 Schematic depiction of the spark plasma sintering apparatus

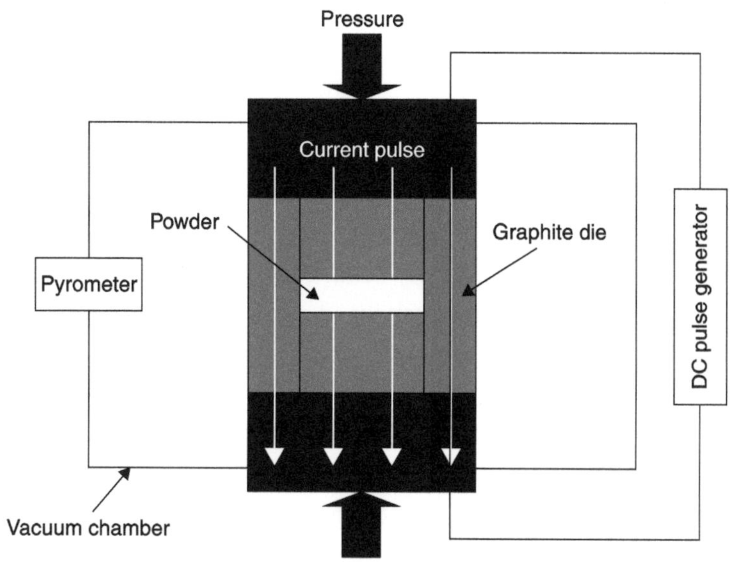

which also facilitates the mass transport. Typical SPS processing parameters include: (a) an applied pressure between 50 and 100 MPa, (b) a pulse duration of ~10 ms with on–off cycle of 2–2.5 ms, and (c) maximum pulse current and tension of 10,000 A and 10 V, respectively.

1.3 Production of nanomaterials by electrodeposition

Electrodeposition is the process of producing a coating, usually metallic, on a surface of a substrate by the action of electric current. The substrate material to be coated is immersed into an electrolyte solution containing a salt of the metal to be deposited and attached to the negative pole of a power supply, i.e. it acts as a cathode in the electrolytic cell [20]. The anode is also immersed into the electrolyte and connected to the positive pole of the power supply. Usually, the anode and the electrolyte are the metal to be deposited and the aqueous solution of its salt, respectively. For example, in the case of Ni the electrolyte may be nickel chloride salt that dissociates in water to positively charged nickel cations and negatively charged chloride anions. As the object to be plated is negatively charged, it attracts the positively charged nickel cations, and electrons flow from the object to the cations to neutralise them to metallic form (see Figure 1.20). Meanwhile, the negatively charged chloride anions are attracted to the positively charged nickel rod. At the anode, electrons are removed from the nickel metal, oxidising it to the nickel cations. Thus, nickel dissolves as ions into the solution.

The deposit formed on the cathode surface may be nanocrystalline, if the electrodeposition parameters, e.g. bath composition, temperature, pH, etc., are properly

Figure 1.20 Schematic illustration of Ni electrodeposition

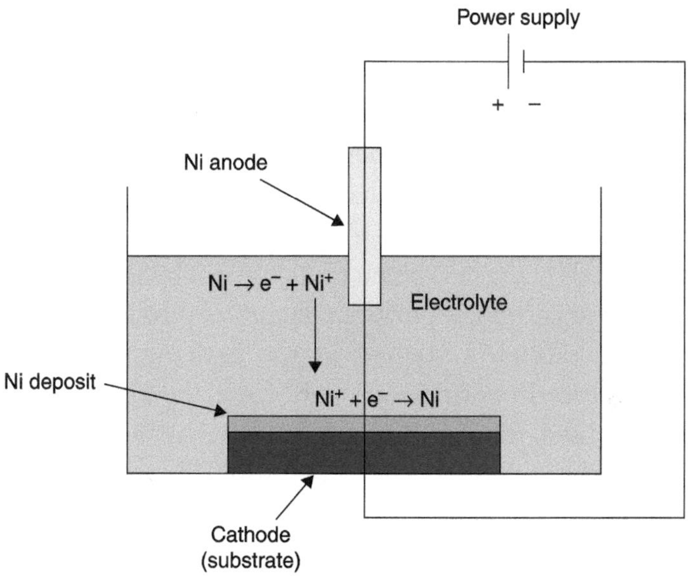

controlled [14]. Pulse plating is particularly attractive because it can yield finer grain structures than that achievable by direct current plating. In pulse plating, the current is imposed in a repetitive square wave with the following controlling parameters: peak current density, pulse-on time and pulse-off time. When the pulse is switched off, the grain growth stops and the next pulse induces nucleation of new grains with other crystallographic orientations, thereby reducing the grain size. The pulse electrodeposition technique permits the application of a very high current density (several orders of magnitude higher than for direct current electrolysis) because the pulse-on time (several tens of milliseconds) is much shorter than the pulse-off time (a few seconds). The metal ion concentration in the vicinity of the cathode, which is greatly decreased during the pulse-on period, can be effectively recovered by ion migration during the relatively long off-time period. In the case of Ni deposition, the bath solution consists of nickel sulphate, nickel chloride, boric acid and saccharin inhibitor ($C_7H_4NO_3S$). The deposits generally have an equiaxed grain structure with fairly narrow grain size distribution. To achieve a significant grain refinement, the pulse-off times must be longer than pulse-on times and a grain refiner such as saccharin is needed to retard the grain growth of Ni deposits. In the absence of saccharin, large crystals in the micrometre range are obtained. However, sulphur and carbon impurities content tend to increase with increasing saccharin content in the bath until saturation occurs. These impurities originate from the chemicals and inhibitor used for the bath [26], and may degradate the mechanical properties of electrodeposits. The electrodeposits are also found to exhibit a texture structure, depending on the bath chemistry. For instance, the preferred orientation of Ni deposits progressively changed from a strong (200) fibre texture for a saccharin-free bath to a (111) texture for a bath containing saccharin [27].

1.4 Nanocrystallisation of bulk amorphous alloys

When a conventional metal or alloy cools from the liquid melt, equilibrium is reached when it solidifies into the lowest energy state structure, i.e. a crystalline lattice. It was discovered that if a molten metal having several components is undercooled uniformly and rapidly enough, the different atoms do not have enough time at the high temperature regime to rearrange for crystal nucleation [28]. The liquid reaches the glass transition temperature, T_g, and solidifies as an amorphous material (metallic glass). Under the glass transition temperature the atomic transport slows down very much, which is also indicated by the large increase of viscosity. At room temperature, the amorphous state remains for a very long time due to the low atomic mobility since crystallisation needs the development of heterogeneous distribution of elements by diffusion. The alloys processable into metallic glasses must have three features: (i) multicomponent systems, (ii) significant atomic size ratios above 12% and (iii) negative heats of mixing [29]. Different rapid cooling processing methods of metallic glasses have been elaborated, e.g. melt-spinning or copper-mould casting. In melt-spinning a wheel is cooled internally, usually by water or liquid nitrogen, and rotated as depicted in Figure 1.21. A thin stream of a molten alloy is then dripped onto the wheel and cooled, causing rapid solidification. The metallic glasses produced by *melt-spinning* have a ribbon shape with small thickness (several tens of micrometres), therefore their practical applications are limited. This technique is used for alloys that require extremely high

Figure 1.21 Illustration of melt-spinning

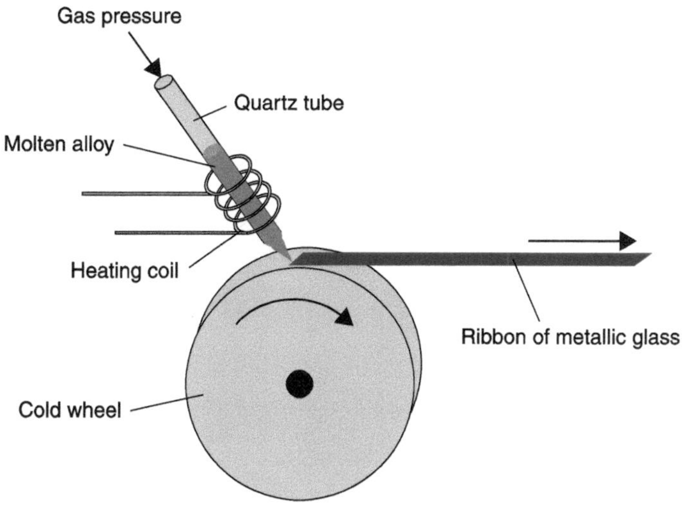

cooling rates in order to form, such as metallic glasses. The cooling rates achievable by melt-spinning are in the order of 10^4–10^7 K/s. In the case of the alloys that can be solidified into an amorphous state even at low cooling rates (1–10 K/s), the *copper-mould casting* technique can be applied for processing metallic glasses with the dimensions of several centimetres. Figure 1.22 shows schematically the copper-mould casting facility. The melt is poured into a water-cooled copper mould under inert atmosphere. The fast flow of the melt into the mould is assisted by a vacuum pump. The melt is stopped in the cooled mould by a copper mesh. The diameter and the length of the as-processed glassy rods are about 10 and 100 mm, respectively. It is noted that bulk amorphous alloys are also produced by hot pressing and warm extrusion of atomised amorphous powders as temperatures fall into the supercooled liquid region.

Careful heat treatment of bulk metallic glasses results in their nanocrystallisation. For multicomponent alloys,

Figure 1.22 Schematic depiction of copper-mould casting

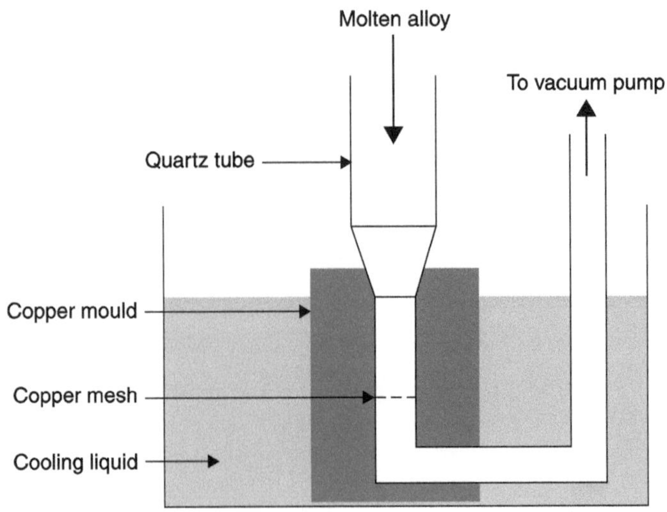

crystallisation usually occurs in two to three successive steps when the time increases at a fixed temperature or the temperature increases at a constant rate during annealing. These steps manifest as distinct exothermic peaks on the thermograms detected by differential scanning calorimetry. In the first step, the metallic glass only partially crystallises and often nanosized, metastable quasicrystalline grains are formed in the amorphous matrix with the volume fraction of about 20–50% [e.g. 30]. It is noted that the redistribution of elements by diffusion is a prerequisite of partial crystallisation or quasicrystallisation, therefore the distribution of nuclei is rather homogeneous. It has been suggested [31, 32] that the ability of metals to form amorphous structure by fast cooling is enhanced by the dominance of icosahedral short-range order (ISRO) in melts, which is incompatible with translational periodicity of crystallographic structures. An icosahedral packing of 20 slightly distorted tetrahedra is more dense than either face-centred cubic (fcc) or hexagonal close-packed (hcp) structures, therefore although it is incompatible with translational periodicity, it might be a natural choice for liquid and amorphous structures [23, 34]. The existence of ISRO in the supercooled liquid state brings about an extremely small interfacial free energy between an icosahedral quasicrystal phase (i-phase) and a metallic glass of the same composition [35]. Consequently, the nucleation of the i-phase during annealing of BMGs is easier than the formation of the more stable crystalline phases. In support of this, the i-phase is frequently reported as the primary devitrification phase, particularly for Zr- and Hf-based bulk metallic glasses [36–38]. It was found that the chemical composition of metallic glasses has a deterministic influence on the local atomic order in the glassy state and therefore on the crystallisation sequence during annealing [39]. Comparing $Zr_{70}Ni_{30}$ and $Zr_{70}Cu_{30}$ metallic glasses, the former shows a tetragonal atomic order while the latter exhibits an icosahedral local atomic configuration. As a consequence, in $Zr_{70}Cu_{30}$ as a first step quasicrystalline i-phase forms while $Zr_{70}Ni_{30}$ crystallises into tetragonal Zr_2Ni phase during heat treatment. It is suggested that structural differences in the glassy phase are caused by a strong chemical affinity of a Zr–Ni pair compared with that of a Zr–Cu pair. The sensitivity of local atomic order to the chemical composition was demonstrated by adding 1 at.% Pd into a $Zr_{70}Al_{10}Ni_{20}$ metallic glass [39]. Without the Pd addition, tetragonal Zr_2Ni was observed in the initial stage of transformation; however, the primary crystallisation process changes markedly into single i-phase formation by the addition of 1 at.% Pd. Addition elements, such as Ag, Pd, Au, Pt, Ti or even oxygen to Zr-based bulk metallic glasses are believed to generate inhomogeneous atomic configuration regions including ISRO configurations in the supercooled liquid [40–44], which then promote the precipitation of an icosahedral phase. In the subsequent crystallisation steps occurring with increasing the time and/or temperature of annealing, stable crystalline phases form from both the quasicrystalline phase and the remaining amorphous matrix. This phase transformation usually occurs at the interfaces of quasicrystalline and amorphous phases (peritectic-type phase transformation) [43].

As an alternative to the annealing of amorphous precursor samples, severe plastic deformation by cold-rolling [45] or HPT [46–48] can also induce nanocrystals at room temperature. As distinct from the uniformly dispersed nanocrystals or nanoquasicrystals formed during annealing, the crystallites induced by cold-rolling are mostly localised in shear bands formed during deformation of metallic glasses [45]. The heterogeneous nucleation of nanocrystals can be attributed to the increase of the free volume at shear bands that facilitates diffusion necessary for crystallisation in these alloys. Additionally, the temperature rises locally at shear bands [49], which also increases atomic mobility. A rather homogeneous dispersion of nanocrystals at an extremely high number density is observed after HPT. For instance, in melt-spun $Al_{88}Y_7Fe_5$ metallic glass processed by five revolutions of HPT, the crystalline particle volume density is 10^{22} m^{-3}, which is larger than the values of 4×10^{21} or 10^{21} m^{-3} obtained after annealing at 245°C for 30 min or by cold-rolling to an equivalent strain of 12, respectively [46]. The maximum size of deformation-induced nanocrystals after cold-rolling [45] and HPT [46] does not exceed 15–18 nm, although the amount of strain varied from about 12 for cold-rolling to about 300 for HPT [46]. The weak dependence of the nanocrystal size on the strain at higher strain levels is explained by a dislocation mediated fragmentation of the deformation-induced nanocrystals exceeding the critical size range [45]. Annealing of the HPT-processed samples yields full nanocrystallisation with a grain size of ~100 nm [46]. In this combined procedure, the HPT treatment gives a very high dispersity of crystalline nuclei, therefore the final grain size is smaller and the size distribution is narrower than that obtained by thermal annealing alone [46].

1.5 References

1. R.Z. Valiev and T.G. Langdon, 'Principles of equal-channel angular pressing as a processing tool for grain refinement', *Progress in Materials Science*, 51 (2006) 881–981.
2. A.P. Zhilyaev and T.G. Langdon, 'Using high-pressure torsion for metal processing: Fundamentals and

applications', *Progress in Materials Science*, 53 (2008) 893–979.

3. V.M. Segal, 'Materials processing by simple shear', *Materials Science and Engineering*, A 197 (1995) 157–164.
4. R.Z. Valiev, R.K. Islamgaliev and I.V. Alexandrov, 'Bulk nanostructured materials from severe plastic deformation', *Progress in Materials Science*, 45 (2000) 103–189.
5. R.Z. Valiev, Y. Estrin, Z. Horita, T.G. Langdon, M. Zehetbauer and Y.T. Zhu, 'Producing bulk ultrafine-grained materials by severe plastic deformation', *Journal of the Minerals, Metals and Materials Society*, 58 (4) (2006) 33–39.
6. Y. Iwahashi, J. Wang, Z. Horita, M. Nemoto and T.G. Langdon, 'Principle of equal-channel angular pressing for the processing of ultra-fine grained materials', *Scripta Materialia*, 35 (1996) 143–146.
7. K. Nakashima, Z. Horita, M. Nemoto and T.G. Langdon, 'Development of a multi-pass facility for equal-channel angular pressing to high total strains', *Materials Science and Engineering*, A 281 (2000) 82–87.
8. T. Watanabe, H. Fujii, H. Oikawa and K.I. Arai, 'Grain boundaries in rapidly solidified and annealed Fe-6.5 mass% Si polycrystalline ribbons with high ductility', *Acta Metallurgica*, 37 (1989) 941–952.
9. G.A. Salischev, S.V. Zherebtsov and R.M. Galeyev, 'Evolution of microstructure andmechanical behavior of titanium during warm multiple deformation' in: Y.T. Zhu, T.G. Langdon, R.S. Mishra, S.L. Semiatin, M.J. Saran and T.C. Lowe (eds), *Ultrafine Grained Materials II*, TMS (The Minerals, Metals and Materials Society), 2002, pp. 123–131.
10. Y. Nakao and H. Miura, 'Nano-grain evolution in austenitic stainless steel during multi-directional

forging', *Materials Science and Engineering*. A 528 (2011) 1310–1317.

11. Y. Beygelzimer, D. Orlov and V. Varyukhin, 'A new severe plastic deformation method: Twist Extrusion' in: Y.T. Zhu, T.G. Langdon, R.S. Mishra, S.L. Semiatin, M.J. Saran and T.C. Lowe (eds), *Ultrafine Grained Materials II*, TMS (The Minerals, Metals and Materials Society), 2002, pp. 297–304.
12. R.A. Varin, T. Czujko and Z.S. Wronski, *Nanomaterials for Solid State Hydrogen Storage*, Springer Science+Business Media, LLC 2009, ISBN 978-0-387-77711-5.
13. C.G. Granqvist and R.A. Buhrman, 'Ultrafine metal particles', *Journal of Applied Physics*, 47 (1976) 2200–2219.
14. S.C. Tjong and H. Chen, 'Nanocrystalline materials and coatings', *Materials Science and Engineering*, R 45 (2004) 1–88.
15. H. Gleiter, 'Nanocrystalline materials', *Progress in Materials Science*, 33 (1989) 223–315.
16. Y. Champion and J. Bigot, 'Synthesis and structural analysis of aluminum nanocrystalline powders', *Nanostructured Materials*, 10 (1998) 1097–1110.
17. Y.A. Kotov, 'Electric explosion of wires as a method for preparation of nanopowders', *Journal of Nanoparticle Research*, 5 (2003) 539–550.
18. W. Kim, J.-S. Park, C.-Y. Suh, H. Chang and J.-C. Lee, 'Fabrication of alloy nanopowders by the electrical explosion of electrodeposited wires', *Materials Letters*, 61 (2007) 4259–4261.
19. A. Alqudami, S. Annapoorni, Govind and S.M. Shivaprasad, 'Ag–Au alloy nanoparticles prepared by electro-exploding wire technique', *Journal of Nanoparticle Research*, 10 (2008) 1027–1036.

20. V. Viswanathan, T. Laha, K. Balani, A. Agarwal and S. Seal, 'Challenges and advances in nanocomposite processing techniques', *Materials Science and Engineering*, R 54 (2006) 121–285.
21. S. Ando, Y. Mine, K. Takashima, S. Itoh and H. Tonda, 'Explosive compaction of Nd-Fe-B powder', *Journal of Materials Processing Technology*, 85 (1999) 142–147.
22. H.V. Atkinson and S. Davies, 'Fundamental aspects of hot isostatic pressing: an overview', *Metallurgical and Materials Transactions*, 31 (2000) 2981–3000.
23. S. Billard, J.P. Fondere, B. Bacroix and G.F. Dirras, 'Macroscopic and microscopic aspects of the deformation and fracture mechanisms of ultrafine-grained aluminum processed by hot isostatic pressing', *Acta Materialia*, 54 (2006) 411–421.
24. H.C. Kim, I.J. Shon, J.E. Garay and Z.A. Munir, 'Consolidation and properties of binderless sub-micron tungsten carbide by field-activated sintering', *International Journal of Refractory Metals and Hard Materials*, 22 (2004) 257–264.
25. G.D. Zhan, J. Kuntz, J. Wan, J. Garay and A.K. Mukherjee, 'A novel processing route to develop a dense nanocrystalline alumina matrix (<100 nm) nanocomposite material', *Journal of the American Ceramic Society*, 86 (2002) 200–202.
26. H. Natter, M. Schmelzer and R. Hempelmann, 'Nanocrystalline nickel and nickel-copper alloys: Synthesis, characterization, and thermal stability', *Journal of Materials Research*, 13 (1998) 1186–1197.
27. A.M. El-Sherik and E. Erb, 'Synthesis of bulk nanocrystalline nickel by pulsed electrodeposition', *Journal of Materials Science*, 30 (1995) 5743–5749.
28. M. Telford, 'The case for bulk metallic glass', *Materials Today*, 7 (2004) 36–43.

29. A. Inoue, 'Stabilization of metallic supercooled liquid and bulk amorphous alloys', *Acta Materialia*, 48 (2000) 279–306.
30. J. Gubicza, J.L. Labar, E. Agocs, D. Fatay and J. Lendvai, 'Effect of nano-quasicrystals on viscosity of a Zr-based bulk metallic glass', *Scripta Materialia*, 58 (2008) 291–294.
31. F.C. Frank, 'Supercooling of liquids', *Proceedings of the Royal Society of London*, A 215 (1952) 43–46.
32. K.F. Kelton, 'Liquid structure and long range diffusion: Their impact on glass formation and nanoscale devitrification', *Intermetallics*, 14 (2006) 966–971.
33. K.F. Kelton, 'Crystallization of liquids and glasses to quasicrystals', *Journal of Non-Crystalline Solids*, 334–335 (2004) 253–258.
34. K.F. Kelton, A.K. Gangopadhyay, T.H. Kim and G.W. Lee, 'A case for local icosahedral order in undercolled metallic liquids and the influence on the nucleation barrier', *Journal of Non-Crystalline Solids*, 352 (2006) 5318–5324.
35. J.C. Holzer and K.F. Kelton, 'Kinetics of the amorphous to icosahedral phase transformation in A–Cu–V alloys', *Acta Metallurgica et Materialia*, 39 (1991) 1833–1843.
36. L.Q. Xing, T.C. Hufnagel, J. Eckert, W. Loser and L. Schultz, 'Relation between short-range order and crystallization behavior in Zr-based amorphous alloys', *Applied Physics Letters*, 77 (2000) 1970–1972.
37. J. Saida, M. Matsushita and A. Inoue 'Nanoicosahedral quasicrystalline phase in Zr–Pd and Zr–Pt binary alloys', *Journal of Applied Physics*, 90 (2001) 4717–4719.
38. H.J. Chang, D.H. Kim, Y.M. Kim, Y.J. Kim and K. Chattopadhyay, 'On the origin of nanocrystals in the shear band in a quasicrystal forming bulk metallic glass

$Ti_{40}Zr_{29}Cu_9Ni_8Be_{14}$', *Scripta Materialia*, 55 (2006) 509–512.

39. J. Saida, M. Imafuku, S. Sato, E. Matsubara and A. Inoue, 'Local structure in quasicrystal-forming Zr-based metallic glasses correlated with a stability of the supercooled liquid state', *Journal of Non-Crystalline Solids*, 353 (2007) 3704–3708.
40. U. Köster, J. Meinhardt, S. Roos and H. Liebertz, 'Formation of quasicrystals in bulk glass forming Zr–Cu–Ni–Al alloys', *Applied Physics Letters*, 69 (1996) 179–181.
41. J. Eckert, N. Mattern, M. Zinkevitch and M. Seidel, 'Crystallization behavior and phase formation in Zr-Al-Cu-Ni metallic glass containing oxygen', *Materials Transactions*, JIM (1998), 39, 623–634.
42. A. Inoue, T. Zhang, J. Saida, M. Matsushita, M.W. Chen and T. Sakurai, 'Formation of Icosahedral Quasicrystalline Phase in Zr-Al-Ni-Cu-M (M = Ag, Pd, Au or Pt) Systems', *Materials Transactions*, JIM 40 (1999) 1181–1184.
43. B.S. Murty and K. Hono, 'Nanoquasicrystallization of Zr-based metallic glasses', *Materials Science and Engineering*, A 312 (2001) 253–261.
44. J. Saida, M. Imafuku, S. Sato, E. Matsubara and A. Inoue, 'Local structure in quasicrystal-forming Zr-based metallic glasses correlated with a stability of the supercooled liquid state', *Journal of Non-Crystalline Solids*, 353 (2007) 3704–3708.
45. R.J. Hebert, J.H. Perepezko, H. Rosner and G. Wilde, 'Dislocation formation during deformation-induced synthesis of nanocrystals in amorphous and partially crystalline amorphous $Al_{88}Y_7Fe_5$ alloy', *Scripta Materialia*, 54 (2006) 25–29.
46. N. Boucharat, R. Hebert, H. Rosner, R. Valiev and G. Wilde, 'Nanocrystallization of amorphous $Al_{88}Y_7Fe_5$

alloy induced by plastic deformation', *Scripta Materialia*, 53 (2005) 823–828.
47. Zs. Kovacs, P. Henits, A.P. Zhilyaev and A. Revesz, 'Deformation induced primary crystallization in a thermally non-primary crystallizing amorphous $Al_{85}Ce_8Ni_5Co_2$ alloy', *Scripta Materialia*, 54 (2006) 1733–1737.
48. P. Henits, A. Revesz, L.K. Varga and Zs. Kovacs, 'The evolution of the microstructure in amorphous $Al_{85}Ce_8Ni_5Co_2$ alloy during heat treatment and severe plastic deformation: A comparative study', *Intermetallics*, 19 (2011) 267–275.
49. J.J. Lewandowski and A.L. Greer, 'Temperature rise at shear bands in metallic glasses', *Nature Materials*, 5 (2006) 15–18.

2

Defect structure in bulk nanomaterials processed by severe plastic deformation

Abstract: Contamination- and porosity-free bulk nanostructured materials can be processed by severe plastic deformation (SPD) of coarse-grained workpieces. During SPD procedures the grain refinement is usually associated with the formation of a large amount of lattice defect, such as grain boundaries, dislocations and vacancies. In this chapter, the evolution of defect structure during SPD-processing is reviewed. At high imposed strain in SPD-processing, the grain and crystallite sizes reach their minimum values and the dislocation density gets saturated. Although there is no strict correlation between the grain or crystallite size and the dislocation density, the higher dislocation density is usually associated with smaller grain and crystallite sizes. The saturation values of the dislocation density and the grain size are strongly influenced by (i) the homologous temperature, the pressure and the strain rate applied in SPD, (ii) the solid solution alloying, (iii) the second phase particles and (iv) the degree of dislocation dissociation. The vacancy concentration in metals processed by SPD at room temperature is as high as the equilibrium value at the melting point (10^{-4}–10^{-3}) and the majority of vacancies are clustered.

Key words: severe plastic deformation, dislocation density, grain refinement, excess vacancies.

2.1 Evolution of dislocation structure and grain size during SPD-processing

Severe plastic deformation (SPD) techniques are effective methods for producing bulk, porosity- and contamination-free ultrafine-grained (UFG) or nanostructured materials [1, 2]. The high pressure applied during these procedures promotes the evolution of a high dislocation density in two ways: (i) by impeding the vacancy migration that hinders the annihilation of dislocations and (ii) by suppressing the cracking, thereby keeping the integrity of the workpiece even at high strains. The most frequently used SPD procedure is equal channel angular pressing (ECAP) that enables the elaboration of bulk UFG or nanomaterials with dimensions of several centimetres in all directions, which is favourable in practical applications [1]. One pass of ECAP corresponds to an equivalent strain value of about 1. The imposed strain increases proportionally by the increase of the number of passes. The evolution of the dislocation density and the crystallite size determined by x-ray line profile analysis as a function of number of ECAP passes is illustrated in Figure 2.1 for 99.98% purity Cu [3, 4], but the tendencies are similar for other metals. The dislocation density increases while the crystallite size decreases with increasing strain in ECAP up to three passes in Figure 2.1. The values obtained after three passes can be regarded as the saturation values achievable by ECAP in pure Cu at room temperature (RT). The saturation of the dislocation density is a consequence of the dynamic equilibrium of the formation and annihilation of dislocations. The maximum dislocation density is reached usually after two to four passes of ECAP for the majority of metallic materials [4–11].

The grain size determined by TEM has similar evolution as for the crystallite size. The minimum grain size in pure Cu processed by ECAP at room temperature is about 200 nm

Figure 2.1 **The dislocation density and the crystallite size as a function of the number of ECAP passes for 99.98% purity Cu**

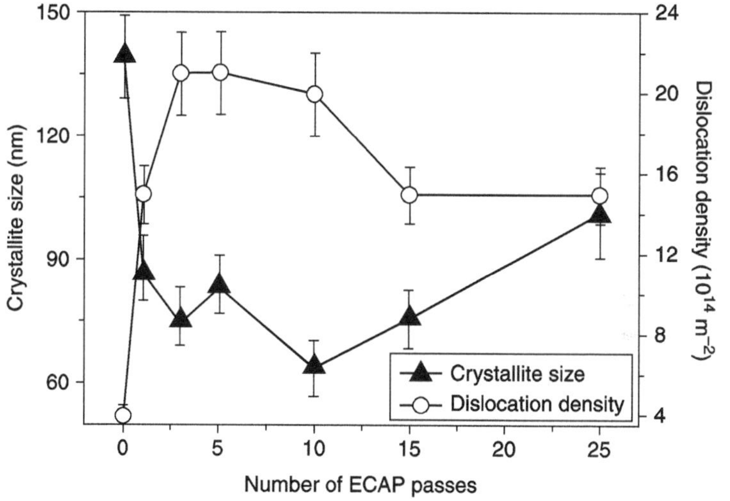

Source: Reprinted from Gubicza et al. (2009) [4] with permission from Elsevier.

[7, 12]. It is noted that the crystallite size measured by x-ray line profile analysis is usually smaller than the grain size obtained by TEM. This phenomenon can be attributed to the fact that the crystallites are the domains in the microstructure which scatter x-rays coherently. As the coherency of x-rays breaks even if they are scattered from volumes having quite small misorientations (1–2°), the crystallite size corresponds rather to the subgrain size in the severely deformed microstructures [13]. The reduction in the subgrain and grain sizes with increasing imposed strain is attributed to the increase of dislocation density as the grain refinement during SPD usually starts by the arrangement of dislocations into low energy configurations, such as low-angle grain boundaries (the misorientation is smaller than 3° per definition). The clustering of dislocations minimises the energy stored in the dislocation structure as the strain field of

dislocations is screened by other dislocations. The clustering of dislocations was observed directly in TEM images [14, 15] and also detected by x-ray line profile analysis [16]. As an example, Figure 2.2 shows a subgrain with a size of about 250 nm in a 99.99% purity Cu sample processed by 14 cycles of repetitive corrugation and straightening (RCS) [14]. The black arrowhead at the left side of the subgrain in Figure 2.2a points to a boundary segment whose Fourier filtered HRTEM image is presented in Figure 2.2b. The inset in Figure 2.2a is an HRTEM image from the subgrain showing that the subgrain boundaries are almost parallel to two sets of {111} planes. The average distance between dislocations in the boundary is about 2 nm as estimated from Figure 2.2b, which yields a local dislocation density of about 3×10^{17} m^{-2} (the average distance between dislocation can be approximated as the inverse square-root of dislocation density). This local dislocation density value is about two orders of magnitude higher than the average dislocation density in the whole sample. In addition to dislocations, other lattice defects such as interstitial and vacancy loops, are also observed in the boundary (see Figure 2.2b).

X-ray line profile analysis also enables the investigation of dislocation clustering as this method gives a dimensionless dislocation arrangement parameter, M, which is defined as the product of the outer cut-off radius of dislocations and the square-root of the dislocation density. Practically, in this quantity the outer cut-off radius of dislocations is normalised by the average dislocation distance that can be expressed as the inverse of the square-root of total dislocation density. The value of M usually decreases with increasing number of ECAP passes [16], which indicates a stronger screening of the strain field of dislocations due to their arrangement into low energy configurations, such as dipolar walls or low-angle grain boundaries.

Figure 2.2 (a) A TEM micrograph of a subgrain in 99.99% purity Cu processed by repetitive corrugation and straightening. The inset in (a) is an HRTEM image showing that the subgrain boundaries are almost parallel to two sets of {111} planes. (b) A Fourier filtered HRTEM image from the boundary as indicated by a black arrowhead in (a). The white arrow in (b) points out the grain boundary orientation. The black and white circles mark interstitial and vacancy loops, respectively

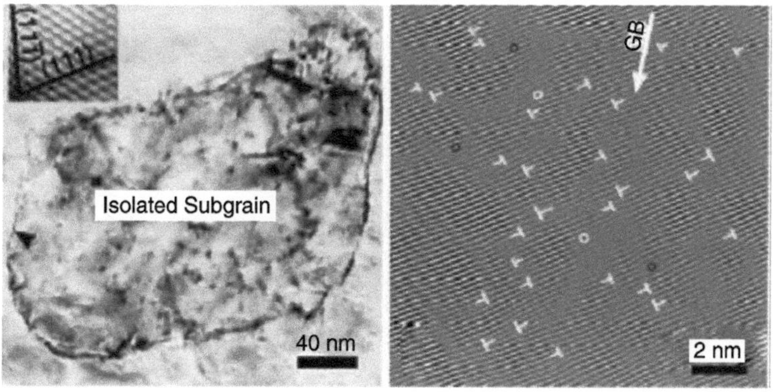

Source: Reprinted from Huang et al. (2001) [14] with permission from Elsevier.

The lattice misorientation across a grain boundary is accommodated by geometrically necessary dislocations. SPD-processing results in non-equilibrium grain boundaries which contain high density of extrinsic dislocations in addition to geometrically necessary dislocations. As an example, Figure 2.3a shows a boundary in a 99.99% purity Cu sample processed by 14 cycles of RCS [14]. The grain boundary plane is curved and changes from the (5 5 12) plane to the (002) plane. The misorientation between the adjacent grains is about 9° as it is revealed by the corresponding electron diffraction pattern in Figure 2.3b.

Figure 2.3 (a) A boundary in a 99.99% purity Cu sample processed by 14 cycles of RCS. The grain boundary plane is curved and changes from the (5 5 12) plane to the (002) plane. (b) The corresponding electron diffraction pattern. (c) and (e) HRTEM images from the upper-left and lower-right part of the boundary in (a) (see the framed areas). (d) A structural model of the boundary segment in (c). The two types of dislocations in the boundary are marked by black and white symbols

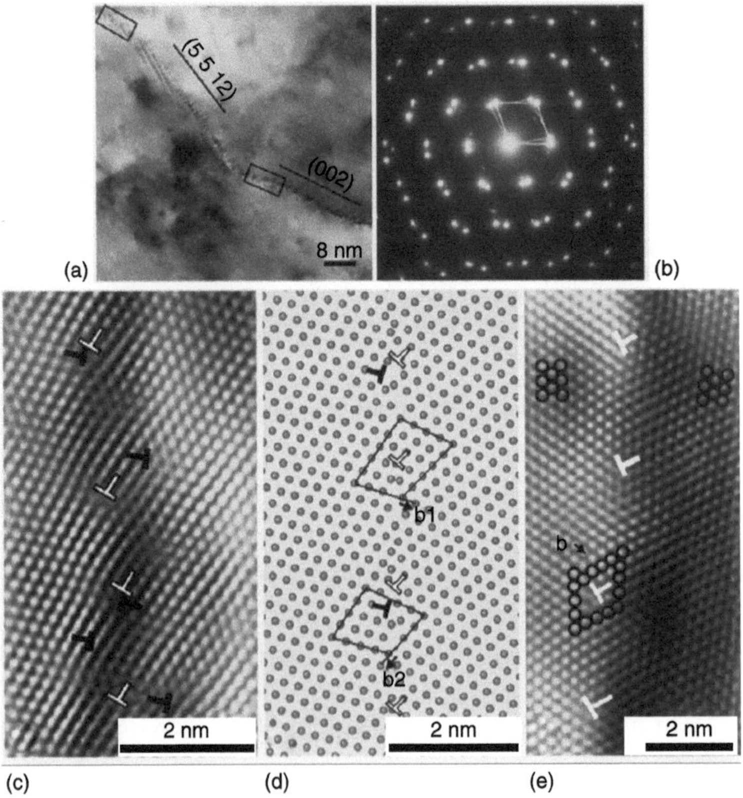

Source: Reprinted from Huang et al. (2001) [14] with permission from Elsevier.

HRTEM images from the upper-left and lower-right part of the boundary (see the framed areas in Figure 2.3a) are shown in Figures 2.3c and 2.3e, respectively. Figure 2.3d is a structural model of the boundary segment in Figure 2.3c. From this model, it can be seen that two types of dislocations marked by black and white symbols are needed to accommodate the geometrical misorientation. These six dislocations are referred to as geometrically necessary or intrinsic dislocations. However, there are three other dislocations in Figure 2.3c which are extrinsic (or non-geometrically necessary) dislocations. Therefore, this segment of grain boundary is in non-equilibrium state. At the same time, the segment in Figure 2.3e contains only geometrically necessary dislocations; therefore that is an equilibrium grain boundary. Large density of extrinsic dislocations in grain boundaries was also observed for hexagonal metals processed by SPD [15].

With increasing strain, the dislocation density in the newly formed subgrain or grain boundaries increases, thereby enhancing the angle of misorientation that can be approximated as the ratio of the magnitude of Burgers vector and the spacing between geometrically necessary dislocations in low-angle grain boundaries. Electron backscatter diffraction (EBSD) experiments revealed that the fraction of high-angle grain boundaries (the misorientation is larger than 15° per definition) monotonously increased with increasing number of ECAP passes even after the achievement of the minimum grain size (about 200 nm for pure Cu) [12]. At very large equivalent strain values ($\varepsilon \approx$ 10–30) of SPD-processing, the fraction of high-angle grain boundaries in various metals (e.g. Cu, Al, Ni, steel) increased up to 70–90% [12, 17–19].

Figure 2.1 shows that between 3 and 10 passes of ECAP, the dislocation density and the crystallite size for Cu remains unchanged within the experimental error. At the same time, during ECAP-processing between 10 and 15 passes the dislocation density decreases and after 25 passes even the crystallite size increases. These changes can be attributed to the structural relaxation of grain/subgrain boundaries. The equilibrium state can be approached by the annihilation of extrinsic dislocations. The decrease of the dislocation density after 15 passes can be explained by this annihilation of extrinsic dislocations. This structural recovery at large strain is accompanied by a decrease in the grain boundary thickness corresponding to an evolution from non-equilibrium boundaries to a more equilibrated structure and by an increase in the misorientation between neighbouring grains. This is illustrated in Figure 2.4, where the TEM images taken on the microstructures processed by 5 and 25 passes are presented [3, 12]. It is noted that the decrease of the dislocation density in SPD-processing can be also observed for other materials (e.g. in Al [6] or Ag [10]) if the imposed strain is high enough and the applied hydrostatic component of the stress field does not suppress recovery processes.

The grain refinement in metallic materials processed by SPD at high homologous temperatures ($0.3–0.4 \times T_m$ or

Figure 2.4 TEM images of the microstructure of Cu processed by ECAP for 5 (a) and 25 (b) passes

Source: Reprinted from Dobatkin et al. (2007) [12] with permission from Elsevier.

above where T_m is the absolute melting point) may occur by continuous dynamic recrystallisation (CDRX) [20]. In CDRX process, new fine grains with HAGBs nucleate homogeneously during SPD as a consequence of the large stored energy. This primary recrystallisation may contribute to grain-refinement in Mg and Al even at room temperature due to their low melting points [21]. A secondary recrystallisation process should be avoided in order to maintain the fine grain structure.

Besides the density and the arrangement of dislocations, x-ray diffraction line profile analysis also enables the determination of the type of the prevailing dislocations in nanomaterials [22]. In the case of cubic crystal systems, the edge/screw character of dislocations can be determined. It was found that for face-centred cubic (fcc) metals having high stacking fault energies (SFEs), e.g. for Al and its alloys, the character of dislocation structure is rather edge type after room temperature SPD-processing [23]. Due to the high SFE, the dissociation of screw dislocations is marginal, which gives a relatively easy annihilation by cross-slip compared with climb of edge dislocation segments. With decreasing SFE, the splitting distance between partials in dissociated dislocations increases, which retards both cross-slip and climb [24, 25]. As cross-slip is more sensitive to the degree of dislocation dissociation than climb [10], the decrease of stacking fault energy leads to a gradual enhancement of the screw character of dislocation structure. In the case of bcc metals, e.g. interstitial free (IF) steel processed by ECAP at room temperature, the character of dislocations is more screw [11], which can be explained by the reduced mobility of screw dislocations compared with edge dislocations in bcc structures. This difficulty in motion of screw dislocations is due to the fact that the ground state dislocation core is dissociated into a non-planar configuration [26]. As a consequence, during ECAP-processing the edge dislocation segments can annihilate more easily than the screw ones, thereby the remaining dislocations have more screw character.

In the case of hexagonal close-packed (hcp) materials, x-ray line profile analysis provides the relative fractions of dislocations having <*a*>, <*c*> and <*c*+*a*> Burgers vectors [27]. Generally, SPD-processing of hexagonal materials is carried out at high homologous temperatures (at about 0.4 × T_m) due to their rigidity. The abundance of <*a*>-type dislocations (usually 60–70%) was found for all hexagonal UFG or nanomaterials (e.g. sintered WC [28, 29], SPD-processed Ti [30, 31] and Mg alloys [32]), which can be explained by their smallest Burgers vector compared with other types of dislocations. The amount of <*c*>-type dislocations was negligible, however a significant fraction (30–40%) of <*c*+*a*> dislocations was found. The relatively high fraction of <*c*+*a*> dislocations in these materials can be attributed to the high temperature of SPD-processing. Theoretical calculations and TEM observations for hexagonal metallic materials suggest the activation of <*c*+*a*> dislocations by strong deformation at elevated temperatures [33]. At room temperature the critical resolved shear stress of pyramidal <*c*+*a*> dislocations is about five times larger than that for basal slip [34], but this value decreases with increasing temperature.

The grain refinement in hcp metals during SPD-processing has different features than in fcc materials. For instance, there is evidence in different Mg-alloys (e.g. AZ31 and ZK60) that above a critical initial coarse grain size, the grain refinement starts with the development of an inhomogeneous grain structure [35–37]. The new grains form along the boundaries of the initial coarse grains in a necklace-like arrangement, as depicted in Figure 2.5. As a consequence, if the initial grain size is large enough, the fine grains at the grain boundaries and the internal volumes of the initial coarse grains give a bimodal grain structure in the beginning

Figure 2.5 **Schematic depiction of the grain refinement in hcp metals along the pre-existing grain boundaries when the initial grain size is larger than a critical value**

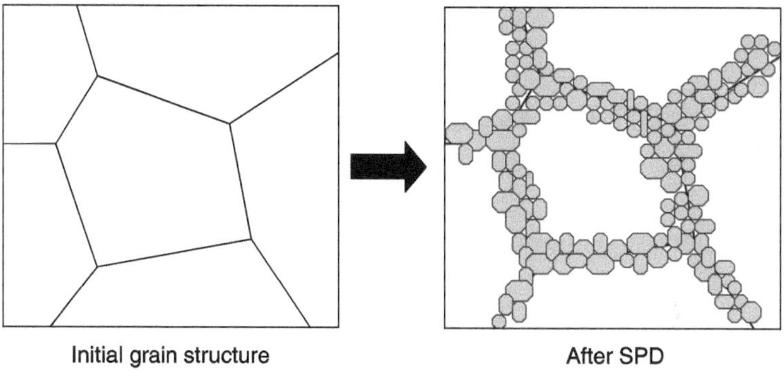

of SPD-processing. With increasing strain, the grain-refinement spreads into the initial grain interiors, thereby leading to a more homogeneous fine grain structure. In hcp metals and alloys, dislocation mechanism of grain refinement requires the activation of both non-basal and basal slips [38]. As dislocations on the basal plane have the smallest <*a*>-type Burgers vector, therefore their activation is easy by the external stresses. The glide on non-basal slip systems is more difficult, and the activation of these dislocations is facilitated by the high stresses and/or the elevated temperature. Therefore, the stress concentrations at the boundaries of the initial grains yield the activation of both basal and non-basal slip processes leading to the formation of fine grains at the pre-existing grain boundaries [39]. If the grain size is smaller than a critical value, the formation of fine grains at the initial grain boundaries does not cause a bimodal grain structure. This critical grain size is between 3 and 9 μm for AZ31 alloy processed by ECAP at room temperature [37]. In this alloy, the bimodal grain size distribution developed during one pass disappears after six passes. However, the critical grain size depends not only on the selected alloy but also on the SPD conditions. For instance, the critical grain size becomes larger when the temperature of ECAP increases, as the higher temperature facilitates the activity of non-basal slip systems.

2.2 Comparison of defect structures formed by different routes of bulk SPD

The effect of SPD-processing method on grain size and the dislocation structure is illustrated on 99.98% purity Cu [40]. The TEM images of the microstructures of Cu samples produced by 20 cycles of MDF, 15 passes of TE, 25 passes of ECAP and 25 revolutions of HPT (at the half-radius) are shown in Figure 2.6. In the TEM images the majority of grains are equiaxed; only a few slightly elongated grains can be observed. The average grain size determined by TEM is 220±5 nm for all Cu samples except the HPT-processed material, which has a smaller grain size of 160 nm. According to the literature data, these values are around the minimum grain size that can be achieved in bulk pure Cu by SPD at RT. The crystallite size, the dislocation density and the twin boundary frequency were determined by x-ray line profile analysis. The dislocation density increases in the order of MDF (7×10^{14} m^{-2}), TE ×10×10^{14} m^{-2}), ECAP (15×10^{14} m^{-2}) and HPT (37×10^{14} m^{-2}) [40]. The higher the dislocation density, the smaller the crystallite size for the four studied samples: 142, 107, 101 and 75 nm for MDF-, TE-, ECAP- and HPT-processed samples, respectively [40]. The twin boundary frequency for all samples was relatively low (0.1±0.1%), close to the detection limit of this quantity for the applied experimental set-up of x-ray line profile analysis. The extremely high dislocation density after HPT is in good agreement with the values obtained by other authors [41].

Figure 2.6 **TEM images showing the microstructure of Cu specimens immediately after 20 cycles of MDF (a), 15 passes of TE (b), 25 passes of ECAP (c) and 25 revolutions of HPT (d)**

Source: Reprinted from Gubicza et al. (2011) [40] with permission from Elsevier.

The very high dislocation density after HPT can be attributed to the high pressure (p = 4 GPa) applied during HPT as it hinders annihilation of dislocations by retarding climb and cross-slip in the following ways:

1. The climb velocity is proportional to the diffusion coefficient [25], which depends on the concentration and the migration enthalpy of vacancies. Previous experiments have shown that in SPD-processed metals a large amount of vacancies are formed due to forced plasticity, therefore their concentration is much higher than the equilibrium value [8, 42–47]. When a high pressure is applied in SPD-processing, larger work (by pV_v where V_v is the volume of

a vacancy) is needed for vacancy formation but this does not reduce the vacancy concentration as the necessary extra work is supplied by the external forces. Rather an increased vacancy concentration was observed in previous studies [42–45] with increasing the pressure of HPT, which can be explained by the suppressed vacancy annihilation at dislocations due to the slower migration of vacancies. The reduction in vacancy migration at higher pressure is caused by the increase of the migration enthalpy by pV_v. The increased difficulty of vacancy migration due to the high pressure applied during HPT hinders effectively the annihilation of dislocations by climb compared with other SPD-processing.

2. In fcc metals the larger the splitting distance between partials in dissociated dislocations, the more difficult the climb or cross-slip of edge or screw segments of dislocations, respectively [24, 25]. The values of equilibrium splitting distance between partials for edge and screw dislocations in Cu are 2.3 and 0.9 nm, respectively [10]. The very large loading on the anvils during HPT most probably results not only in a high hydrostatic pressure but also in high shear stresses since the idealised constrained conditions are usually not achieved. The shear stresses acting on the slip plane of a dislocation perpendicular to the Burgers vector can decrease or increase the splitting distance by pushing the partials towards each other or pulling the partials in opposite directions, respectively, depending on the stress orientation (see Figure 3.2 in Chapter 3). Therefore, the higher shear stresses during HPT may result in much larger or smaller splitting distance depending on their directions. As the splitting distance without stresses is not very large in Cu due to the medium value of SFE (data between 45 and 78 mJ/m^2 are reported in the literature

[48, 49]), the stresses pushing the partials towards each other do not accelerate climb and cross-slip very much. At the same time, the high stresses resulting in much larger splitting distance slow down climb and cross-slip effectively. As a result of the effects (1) and (2), both climb and cross-slip of dislocations are retarded due to the high load applied during HPT, which explains the extremely large dislocation density measured immediately after HPT-processing. This argumentation is also supported by the increase of dislocation density with increasing pressure applied during HPT [e.g. 42, 47].

The grain boundary character depends on both the imposed strain and the route of SPD-processing. The evolution of a reasonably equiaxed ultrafine-grained microstructure with large fraction of high-angle grain boundaries is usually most rapid for route B_C among the different routes of ECAP [50, 51], however the difference in the grain boundary misorientation distribution produced by the various ECAP routes decreases with increasing number of passes [12]. At very high imposed strains ($\varepsilon \approx 10$–40), the high-angle grain boundary fraction is the largest in the materials processed by ECAP and HPT compared with other SPD methods [12, 17].

2.3 Maximum dislocation density and minimum grain size achievable by SPD of bulk metallic materials

At high imposed strain in SPD-processing, the grain size and the crystallite size (subgrain size) reach their minimum values and the dislocation density gets saturated. The saturation dislocation density, crystallite and grain sizes for different SPD-processed metals are listed in Table 2.1. In the case of

Table 2.1 The maximum dislocation density and the minimum crystallite size determined by x-ray line profile analysis, and the minimum grain size obtained by TEM for metallic materials processed by SPD. RT: room temperature

Material	Processing method	Grain size [nm]	Crystallite size [nm]	Dislocation density [10^{14} m^{-2}]	Reference
Pure metals					
Al	ECAP at RT	1,200	272	1.8	[6]
Al	HPT at RT	800	–	–	[53]
Al	Cryogenic rolling at 77 K, reduction = 0.75	200–300	–	–	[54]
Cu	ECAP at RT	200–220	64–75	21–26	[4, 7]
Cu	HPT at RT	160	75	37–40	[40, 45]
Au	ECAP at RT	460	72	17	[55]
Ni	8 ECAP + 5 HPT at RT	140	48	25	[56]
Ni	HPT at RT	–	–	34	[45]
Ag	ECAP at RT	200	60	46	[10]
Ti	ECAP at 700 K	265	40	30	[15]
Solid solution alloys					
Al-1Mg	ECAP at RT	450	88	3.9	[9]
Al-3Mg	ECAP at RT	300	65	23	[9]

Al-7075, solid solution	HPT at RT	26	–	320	[57]
IF Steel	ECAP at RT	360	66	10	[11]
Cu-10%Zn	HPT at RT	145	60	34	[52]
Cu-30%Zn	HPT at RT	85	34	81	[52]
Cu-5% Al	ECAP at RT	107	–	–	[58]
Cu-8% Al	ECAP at RT	82	–	–	[58]
Dispersion-strengthened alloys					
Al 6082	ECAP at RT	300	76	5.4	[9]
Al-4.8Zn-1.2Mg-0.14Zr	ECAP at 473 K	500	165	3.2	[59]
Al-5.7Zn-1.9Mg-0.35Cu	ECAP at 473 K	300	119	3.4	[59]
Al-5.9Mg-0.3Sc-0.18Zr	HPT at RT	50	37	24	[13]
Mg-9Al-1Zn-0.2Mn	ECAP at 543 K	1,200	97	2.0	[32]
Cu-0.18%Zr	HPT at RT	280	100	20	[60]
Cu-0.18%Zr	ECAP at RT	250	73	39	[61]

HPT-processed materials, the data measured at the half-radius of the disks are selected as characteristic values. For Cu-10%Zn and Cu-30%Zn alloys [52], the grain size values at the half-radius are not available; therefore the mean of the data measured in the centre and the periphery is presented in Table 2.1. The minimum grain and crystallite size values as a function of the maximum dislocation density are plotted in Figure 2.7. It can be concluded that, although there is no strict correlation between the grain or crystallite size and the dislocation density of the UFG metals processed by SPD

Figure 2.7 The saturation grain and crystallite size values determined by TEM and x-ray line profile analysis, respectively, for SPD-processed UFG materials as a function of the saturation dislocation density. The data are taken from Table 2.1 and plotted in double logarithmic scales as the values span about two orders of magnitude. The experimental error corresponds to the size of symbols

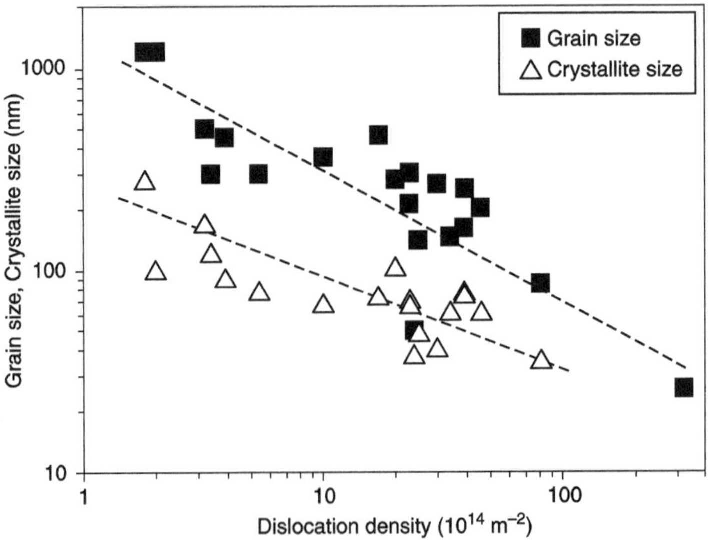

procedures, the higher dislocation density is associated with smaller grain and crystallite sizes. This is a consequence of the fact that the basic mechanism of grain-refinement in SPD materials is the arrangement of dislocations into grain and subgrain boundaries. Table 2.1 also shows that the grain size determined by TEM is usually two to six times larger than the crystallite size obtained by x-ray line profile analysis. This is a consequence of the fact that the coherently scattering domains (crystallites) in SPD-processed materials correspond to the subgrains and/or the dislocation cells inside the grains bounded by high-angle grain boundaries.

The saturation dislocation density is determined by the dynamic equilibrium between the formation and annihilation of dislocations. The more difficult the dislocation annihilation during SPD-processing, the higher the maximum dislocation density. The annihilation processes are hindered by (i) the low homologous temperature of SPD, (ii) the solid solution alloying, (iii) the second phase particles and (iv) the high degree of dislocation dissociation due to the low SFE. The homologous temperature of SPD can be kept at low values either by processing at low temperatures (e.g. cryogenic rolling at liquid nitrogen temperature, 77 K [54]) or by applying SPD on metals having high melting points. Figure 2.8 illustrates that the higher the melting point (T_m) in pure fcc metals having medium or high SFE (Al, Ni, Cu and Au), the larger the maximum dislocation density achievable by ECAP at room temperature. As the grain refinement occurs by arranging of dislocations into subgrain and grain boundaries, the higher saturation dislocation density is accompanied by smaller crystallite and grain sizes, as shown in Figure 2.8. The effect of the melting point on the saturation dislocation density can be explained by the thermally activated processes (cross-slip and climb) of dislocation annihilation. For instance, in fcc metals having medium or

Figure 2.8 The saturation crystallite size and dislocation density as a function of melting point for different pure fcc metals processed by ECAP at RT

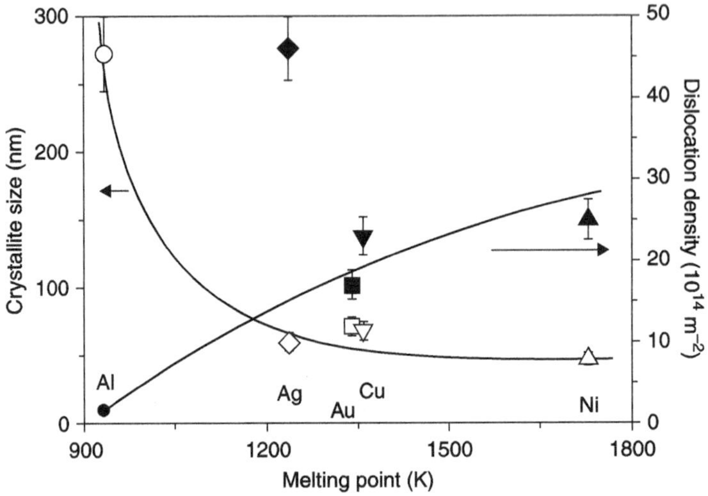

high SFE, the degree of dislocation dissociation is not very large, therefore cross-slip of screw dislocations occurs more easily than climb of edge segments during SPD-processing. As a consequence, the maximum dislocation density at high strains is determined by the diffusion-controlled climb process. The diffusion coefficient is given by:

$$D = D_0 exp \frac{-Q_{self}}{RT}, \tag{2.1}$$

where Q_{self} is the activation energy of self-diffusion, R is the universal gas-constant and T is the absolute temperature. It is known that Q_{self} for fcc metals is proportional to the melting point [62]:

$$Q_{self} = B \cdot T_m, \tag{2.2}$$

where $B \approx 155\ J \cdot K^{-1} \cdot mole^{-1}$. As the pre-exponential factor, D_0 has similar values for Al, Cu, Au and Ni, the diffusion coefficient depends primarily on the homologous temperature (T/T_m) due to Equations 2.1 and 2.2. At room temperature, the larger the melting point, the higher the activation energy and therefore the slower the climb process resulting in larger saturation dislocation density.

It is noted that climb is usually considered as an important process only at high temperatures due to the slow diffusion at RT. However, UFG materials produced by SPD have a high volume fraction of grain-boundaries and a large density of dislocations together with a large concentration of excess vacancies. The activation energy of diffusion along grain boundaries and dislocation cores is about half of that for self-diffusion, therefore the vacancy migration in nanomaterials is faster than in their coarse-grained counterparts by a factor of 10^{10} to 10^{20} [63]. Consequently, diffusion may control the annihilation of dislocations even at room temperature. The high internal stresses developed in SPD-processed nanomaterials due to the large defect densities also assist climbing considerably.

Figure 2.8 shows that the saturation dislocation density for pure Ag is much higher than the values for other fcc metals having similar melting points (Au, Cu). The very large dislocation density in Ag can be explained by its extremely low SFE (16 mJ/m^2 [48]) yielding a high degree of dislocation dissociation, which hinders both cross-slip and climb, thereby retarding dislocation annihilation. The evolution of the defect structure in low SFE materials will be presented in Chapter 3.

The temperature and strain rate of SPD-processing have a significant effect on the defect structure and grain size in nanocrystalline or ultrafine-grained metals and alloys. For a given strain and processing method, the higher the strain rate and/or the lower the temperature of deformation, the larger the dislocation density and smaller the grain size. The combined effects of strain rate ($\dot{\varepsilon}$) and temperature can be expressed by the Zener-Hollomon parameter, defined as:

$$Z = \dot{\varepsilon}\, exp\left(\frac{Q}{RT}\right), \tag{2.3}$$

where Q is the related activation energy for the dominant diffusion process (bulk or grain boundary diffusion). For a constant strain value, the following relationship between the grain size (d) and parameter Z is suggested:

$$ln\, d = A - B\, ln\, Z, \tag{2.4}$$

where A and B depend on the studied material. In the case of Cu, assuming a grain boundary diffusion activation energy of 73 kJ/mol, the value of lnZ is 22 for room temperature deformation at a strain rate of 10^{-3} s^{-1} [64]. With increasing lnZ from 22 to 66 in uniaxial deformation of Cu, the grain size decreases from 320 to 66 nm, while the dislocation density increases from 14×10^{14} m^{-2} to 30×10^{14} m^{-2} at a fixed equivalent strain of 2 [64]. Deformation twinning occurs in Cu when lnZ exceeds 30, and the twin boundary frequency increases at higher Z. The grain sizes achieved in Cu at different strain rates and temperatures of SPD are close to each other for similar values of parameter Z [65].

Similar to pure metals, in solid solution alloys the dislocation density, the crystallite and grain sizes saturate after ~4–8 passes of ECAP [5–9, 66–69]. At the same time, alloying usually leads to an increase of the critical imposed strain corresponding to saturation. For instance, in 99.99% purity Al and Al–3%Mg alloy processed by ECAP at RT, the maximum value of the dislocation density is achieved after one and four passes, respectively. Additionally, the maximum dislocation density increases while the minimum values of the crystallite and grain sizes decrease due to alloying. As an example, Figure 2.9 shows the saturation values of the dislocation density and the grain size as a function of the Mg concentration in Al processed by ECAP at RT [9]. Both the

Figure 2.9 **The minimum grain size and the maximum dislocation density as a function of Mg content in Al solid solutions**

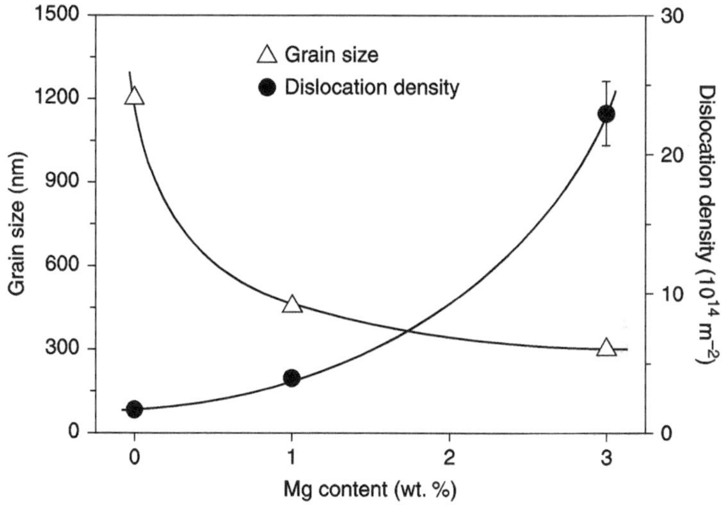

Source: Reprinted from Gubicza et al. (2006) [9] with permission from Elsevier.

larger strain needed for saturation and the higher value of the saturation dislocation density in alloys can be explained by the pinning effect of solute atoms on dislocations, which hinders their annihilation. As grain refinement in SPD metals occurs usually by the re-arrangement of dislocations into subgrain and grain boundaries, the higher dislocation density results in a decrease of grain size for higher Mg concentrations as shown in Figure 2.9 and demonstrated in the TEM images of Figure 2.10. In pure Al the saturation grain size is 1,200 nm while 3 wt.% Mg alloying leads to a reduction to 300 nm. It is noted that although the minimum grain size in Al–3% Mg alloy is only four times smaller than in pure Al, the maximum dislocation density is approximately 13 times higher (see Table 2.1). The dragging effect of solute atoms on dislocations also hinders the evolution of low-angle grain boundaries into high-angle boundaries during SPD. For instance, although in both

Figure 2.10 **TEM images taken on pure Al (a) and Al–3%Mg (b) processed by 8 ECAP passes at room temperature**

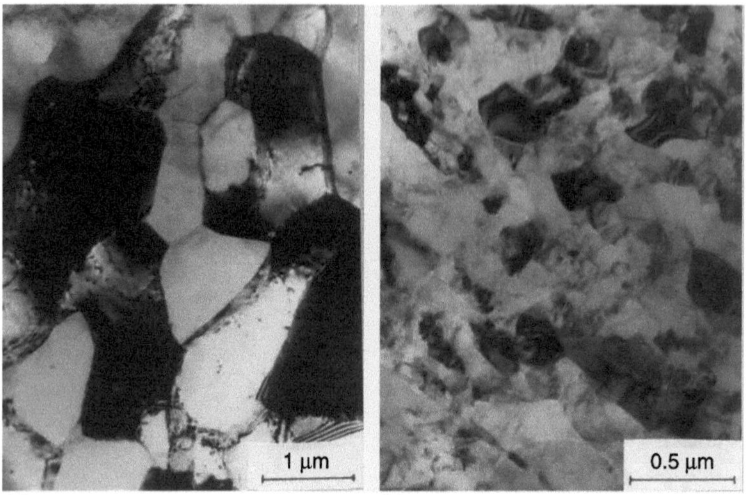

(a) (b)

Source: Reprinted from Gubicza et al. (2004) [6] with permission from Elsevier.

99.99% purity Al and Al–1% Mg alloy the fraction of high-angle grain boundaries increases with increasing the number of ECAP passes, its maximum value in Al–1% Mg (~65%) is lower than that for pure Al (~75%) [70, 71].

When SPD-processing is performed on supersaturated solid solution alloys at high temperatures, grain refinement and formation of precipitates occur simultaneously. The precipitation processes are promoted by SPD since dislocations act as nucleation sites for precipitates. This can be explained by the faster diffusion of alloying elements within the cores of the dislocations and by the larger space for the alloying atoms in the dilatational zones around the dislocations. For instance, in Al–4.8Zn-1.2Mg-0.14Zr and Al–5.7Zn-1.9Mg-0.35Cu (wt.%) alloys processed by eight passes of ECAP at 200°C stable incoherent $MgZn_2$

precipitates (η-phase particles) form while artificial ageing at the same temperature and for the same time (30 minutes) as for ECAP yields only negligible concentration of η precipitates and mainly coherent Guinier-Preston zones develop [59]. The experimental results show that processing by ECAP at a high temperature not only influences the precipitation kinetics but also there is a significant effect on the shape of the η precipitates. The Al-Zn-Mg-Zr and Al-Zn-Mg-Cu alloys aged for long times contain long rod-like precipitates but these are absent and only essentially spherical particles are present in the specimens processed by ECAP [59]. On the other hand, the precipitation during SPD most probably influences the grain refinement since the pinning effect of the second phase particles on dislocations and grain boundaries yields higher dislocation density and smaller grain size than for the pure counterparts. In Al-Zn-Mg-Zr and Al-Zn-Mg-Cu alloys, ECAP-processing for eight passes results in a two times higher dislocation density (3.2–3.4 × 10^{14} m^{-2}) and a three times smaller grain size (300–500 nm) than in pure Al, despite the much higher temperature of ECAP in the former case (200°C for the alloy and RT for pure Al) [59].

It is worthwhile mentioning that the minimum grain size achievable by SPD is rather affected by the deformation conditions than the initial grain size before deformation. For instance, if room temperature HPT-processing is performed on electrodeposited nanostructured Ni with the grain size of 30 nm, there is a grain-growth to 129 nm [72], which is close to the grain size limit for HPT-treated coarse-grained Ni (140–170 nm [18]). Although the dislocation density increases from 120 × 10^{14} to 200 × 10^{14} m^{-2} during HPT of electrodeposited Ni, this cannot compensate the softening effect of grain growth and the hardness decreases from 7.2 to 6.1 GPa. Similar grain growth was also observed during indentation of nanomaterials, while there is a lack of grain growth during tensile testing.

This suggests that only stresses above a critical value induce grain growth in nanocrystalline materials. The large dislocation density in the electrodeposited Ni after HPT may be a consequence of the high impurity content that is usually a characteristic feature of electrodeposited materials.

2.4 Excess vacancy concentration due to SPD

It is well known that during deformation in addition to dislocations, vacancies are also formed. The most important phenomenon resulting in vacancy production is the climb of edge dislocation segments. However, an edge dislocation may climb to different directions relative to its extra half-plane, therefore it may act as either a source or a sink of vacancies. Figures 2.11a and 2.11b show schematic drawings of two edge dislocations climbing in opposite directions by increasing or decreasing the extension of their extra half-planes, which leads to a production or annihilation of vacancies, respectively. The significant difference between the two climb processes is that the second one needs vacancy migration to the core of dislocation contrary to the first one. As a consequence, in the beginning of SPD-processing at RT, due to the slow diffusion the climb-producing vacancy occurs more easily than the climb-sinking vacancy. Therefore, the vacancy production rate is higher than the rate of annihilation, resulting in an increase of excess vacancy concentration with increasing strain. At large strain values, the vacancy migration becomes faster due to the large amounts of dislocations and grain boundaries, therefore the formation and annihilation of vacancies equilibrate, yielding to a saturation concentration of excess vacancies at the given conditions of SPD.

Figure 2.11 Schematic drawings of two edge dislocations climbing in opposite directions by increasing or decreasing the extension of the extra half-plane, which leads to a production or annihilation of vacancies, respectively. The direction of atomic migration is indicated by arrows. The produced and the annihilated vacancies are denoted by solid and dashed squares, respectively

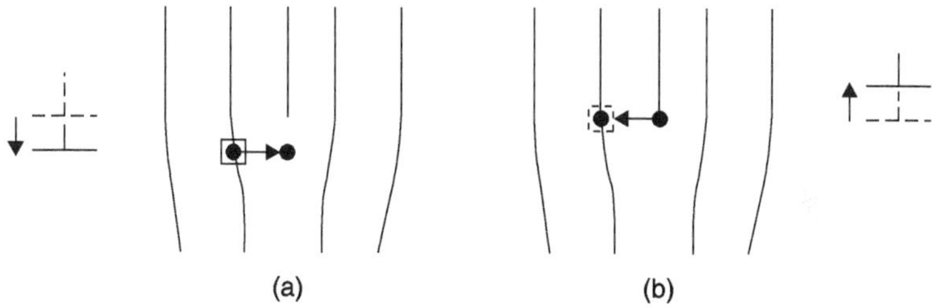

The concentration of excess vacancies was investigated mainly in Cu and Ni processed by ECAP and HPT [8, 45, 62–68]. In HPT-deformed 99.95% purity Cu, positron annihilation spectroscopy (PAS) revealed the absence of monovacancies, while a large amount of vacancy clusters were identified [73]. Most probably, the monovacancies created during HPT-processing either disappear by diffusion to sinks such as grain boundaries or agglomerate into small clusters. The vacancy clusters at the centre of the HPT-processed Cu and Ni disks consist of 4–5 vacancies, while the clusters at the periphery consist of 7–9 vacancies [73]. Neither the increase of number of HPT revolutions nor the rise in pressure from 2 to 4 GPa results in a change in the size of vacancy clusters. The densities of dislocations and vacancies were so high in the HPT-processed Cu samples that every positron was very quickly trapped at dislocations or vacancy clusters (referred to as saturated positron trapping). As a consequence, the concentration of vacancy clusters could only be evaluated from the PAS signal with the help of the dislocation density determined by x-ray line profile analysis. The ratio, I_2/I_1, of the intensities of PAS signals corresponding to positrons trapped in vacancy clusters and at dislocations is directly proportional to the ratio of cluster concentration and dislocation density [73]:

$$\frac{I_2}{I_1} = \frac{\nu_V c_V}{\nu_D \rho}, \tag{2.5}$$

where ν_V, ν_D are the specific positron trapping rate to vacancy clusters and dislocations, respectively, c_V is the vacancy cluster concentration and ρ is the dislocation density. The values of ν_V, ν_D for Cu are 1.2×10^{14} at s^{-1} and 0.6×10^{-4} m^2 s^{-1}, respectively [73]. This methodology most probably overestimates the vacancy cluster concentration as the calculation uses the total dislocation density while the positrons are mainly trapped at edge dislocation segments. The concentration of vacancy clusters increases with increasing of number of HPT revolutions in Cu from about 1.0×10^{-4} after 1 turn to about 2.5×10^{-4} after 25 revolutions, as is shown in Figure 2.12 [73]. Multiplying the cluster concentration with the numbers of vacancies in the clusters, the total vacancy concentration is determined for the centre and the periphery of the disks as a function of number of revolutions and plotted in Figure 2.13 [73]. The periphery exhibits enhanced concentration of vacancies due to larger size of vacancy clusters. The concentration of vacancies at both the centre and the periphery of the disks increase with increasing number of HPT revolutions due to the increase of the imposed strain. At the centre, the vacancy concentration is 5×10^{-4} after 1 turn, which increases to 12×10^{-4} after 25 revolutions. At the periphery, the corresponding concentrations are 12×10^{-4} and 20×10^{-4} after 1 and 25 revolutions, respectively. These values have the same order of magnitude as the concentrations observed in other Cu and

Figure 2.12 The vacancy cluster concentration in HPT-processed Cu as a function of the number of revolutions. The individual datum points given in [73] are not shown here, but all of them follow this line irrespectively of the location in the HPT disk and the applied pressure (2 or 4GPa). The scatter of datum points is illustrated by the error bar

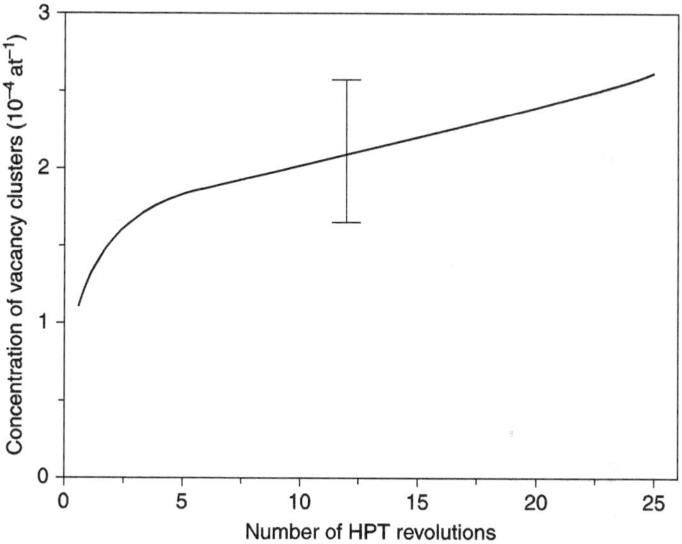

Ni disks processed by HPT at RT [45]. The vacancy concentration and the cluster size determined for different SPD-processed samples are listed in Table 2.2.

For ECAP-processed Cu, the vacancy concentration was determined by a combination of x-ray line profile analysis and residual electrical resistometry (RER) or differential scanning calorimetry (DSC) [8]. The decrease of electrical resistivity during annealing of the ECAP-processed samples is caused by the annihilation of lattice defects such as vacancies (vacancy clusters) and dislocations. The contribution of dislocations to resistivity can be estimated from the dislocation density determined by x-ray line profile

Figure 2.13 The concentration of vacancies in the centre and at the periphery of Cu disks processed by different revolutions of HPT. The individual datum points given in [73] are not shown here, but all of them follow this line irrespectively of the applied pressure (2 or 4GPa). The scatter of datum points is illustrated by the error bar

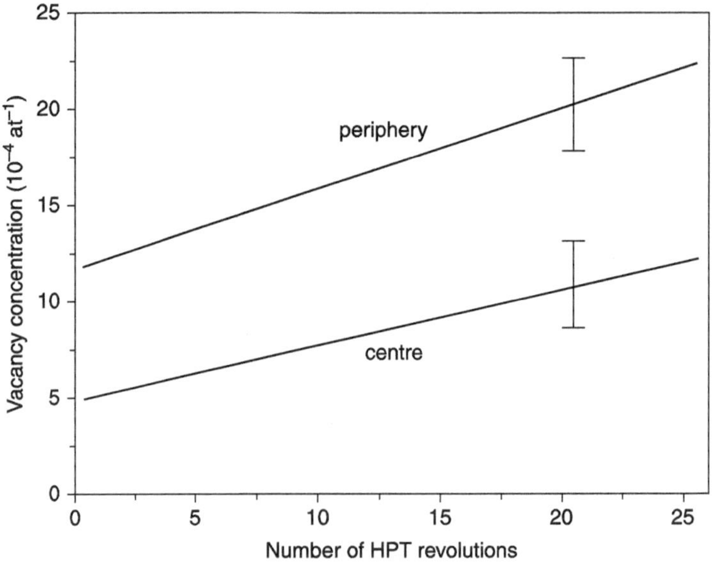

analysis and the specific dislocation resistivity of Cu being 0.8×10^{-25} Ωm^3 [74]. Subtracting this contribution from the total decrease of resistivity, the vacancy concentration can be determined as the ratio of the remaining resistivity and the resistivity per unit vacancy concentration (0.62×10^{-4} Ωcm for Cu [75]). The vacancy concentrations obtained in this way are 1.5×10^{-4} and 3.5×10^{-4} after one and four ECAP passes, respectively. These values are smaller than the concentrations determined in HPT-processed Cu samples. This can be explained by the larger imposed strain in HPT compared with ECAP. The vacancy concentration values determined from the released heat measured in DSC agree well with those obtained

Table 2.2 The concentration and the cluster size of vacancies obtained for metals processed by SPD. RT: room temperature

Material	Processing method	Cluster size	Vacancy concentration	Reference
Cu	1 ECAP at RT	–	1.5×10^{-4}	[8]
Cu	4 ECAP at RT	–	3.5×10^{-4}	[8]
99.95% Cu	1 HPT at RT, centre of disk, p=2–4 GPa	4–5	5×10^{-4}	[73]
99.95% Cu	1 HPT at RT, periphery of disk, p=2–4 GPa	7–9	12×10^{-4}	[73]
99.95% Cu	25 HPT at RT, centre of disk, p=2–4 GPa	4–5	12×10^{-4}	[73]
99.95% Cu	25 HPT at RT, periphery of disk, p=2–4 GPa	7–9	20×10^{-4}	[73]
99.99% Cu	HPT at RT, p=4–8 GPa, shear strain=20	–	4×10^{-4}	[45]
99.998% Ni	HPT at RT, p=2 GPa, shear strain=20	1–2+ clusters with unknown size	1.1×10^{-4} (single/ double vacancies) 3.5×10^{-4} (in clusters)	[45]
99.998% Ni	HPT at RT, p=8 GPa, shear strain=20	1–2+ clusters with unknown size	0.9×10^{-4} (single/ double vacancies) 5×10^{-4} (in clusters)	[45]
Cu-0.18%Zr	8 ECAP at RT	4	5×10^{-5}	[61]

by resistometry [8]. In the DSC thermograms for ECAP-processed Cu samples, one exothermic peak was observed. The released heat is assumed to be a sum of the contributions of vacancies and dislocations. The stored energy in a unit volume corresponding to dislocations having half edge – half screw character is determined as [8]:

$$E_{disl} = 0.2Gb^2 \rho \, ln \frac{1}{b\sqrt{\rho}}, \tag{2.6}$$

where G is the shear modulus (47 GPa for Cu), b is the magnitude of Burgers vector (0.25 nm for Cu) and ρ is the dislocation density. Subtracting E_{disl} from the released heat and dividing the remaining stored energy by the formation energy of a vacancy, 0.195×10^{-18} J [75], the vacancy concentration can be determined. It is worth noting that the values of c_v obtained for Cu after ECAP or HPT at RT has the same order of magnitude as the equilibrium vacancy concentration near the melting point. Taking 1.28 eV as the value of the vacancy formation energy (E_F) and 12.2 as the pre-exponential factor (c_0) [76], the Arrhenius-type formula of vacancy concentration:

$$c_V = c_0 \, exp \frac{-E_F}{RT}, \tag{2.7}$$

gives 4×10^{-21} and 2×10^{-4} for the equilibrium vacancy concentrations at RT and the melting point, respectively. This means that the vacancy concentration after ECAP at RT is 17 orders of magnitude larger than its equilibrium value due to severe plastic deformation. It is noted that vacancies have a considerable contribution to the diffuse background scattering under x-ray diffraction peaks, therefore the ratio of the integrated background and the integrated peak intensity can be used for the determination of vacancy concentration after a calibration procedure [77].

In the case of HPT-processed 99.99% and 99.998% purity Ni, two exothermic peaks were observed on the DSC thermograms [45]. The first peak at lower temperature corresponds to the disappearance of single/double vacancies while the second one at higher temperature is related to the annihilation of vacancy clusters and dislocations. The first DSC peak was not observed for 99.99% Cu processed under the same conditions (see above) as most probably all the vacancies are clustered due to the smaller values of stacking fault energy and melting point of Cu. In Ni with lower purity than 99.99%, single/double vacancy peak is also not detected as impurities facilitate the formation of vacancy clusters. The single/double vacancy concentration in Ni was determined as the ratio of the heat released in the first peak and the formation energy of a vacancy (0.29×10^{-18} J for Ni [75]). The concentration of single/double vacancies increases with increasing shear strain during HPT and it saturates at about $1.0 \pm 0.1 \times 10^{-4}$ after a shear strain of about 20 irrespectively of the pressure (2–8 GPa) [45]. The concentration of vacancy clusters also increases with increasing strain and its saturation values are $3.5 \pm 1.0 \times 10^{-4}$ and $5.1 \pm 1.0 \times 10^{-4}$ for 2 and 8 GPa, respectively [45]. The cluster size in the HPT-processed Ni samples was not determined. However, if we assume that the clusters consist of 4–9 vacancies similarly as in Cu, and we take into account the difference between the concentrations of clusters and single/double vacancies, it is evident that more than 90% of vacancies are clustered.

2.5 References

1. R.Z. Valiev and T.G. Langdon, 'Principles of equal-channel angular pressing as a processing tool for grain

refinement', *Progress in Materials Science*, 51 (2006) 881–981.

2. A.P. Zhilyaev and T.G. Langdon, 'Using high-pressure torsion for metal processing: Fundamentals and applications', *Progress in Materials Science*, 53 (2008) 893–979.
3. J. Gubicza, S. Dobatkin, Z. Bakai, N.Q. Chinh and T.G. Langdon, 'Microstructure and Mechanical Behavior of Severely Deformed F.C.C. Metals', *Materials Science Forum*, 567–568 (2007) 181–184.
4. J. Gubicza, N.Q. Chinh, J.L. Lábár, S. Dobatkin, Z. Hegedűs and T.G. Langdon, 'Correlation between microstructure and mechanical properties of severely deformed metals', *Journal of Alloys and Compounds*, 483 (2009) 271–274.
5. F. Dalla Torre, R. Lapovok, J. Sandlin, P.F. Thomson, C.H.J. Davies and E.V. Pereloma, 'Microstructures and properties of copper processed by equal channel angular extrusion for 1–16 passes', *Acta Materialia*, 52 (2004) 4819–4832.
6. J. Gubicza, N.Q. Chinh, Z. Horita and T.G. Langdon, 'Effect of Mg addition on microstructure and mechanical properties of aluminum', *Materials Science and Engineering*, A 387–389 (2004) 55–59.
7. J. Gubicza, L. Balogh, R.J. Hellmig, Y. Estrin and T. Ungár, 'Dislocation structure and crystallite size in severely deformed copper by X-ray peak profile analysis', *Materials Science and Engineering*, A 400–401 (2005) 334–338.
8. E. Schafler, G. Steiner, E. Korznikova, M. Kerber and M.J. Zehetbauer, 'Lattice defect investigation of ECAP-Cu by means of X-ray line profile analysis, calorimetry and electrical resistometry', *Materials Science and Engineering*, A 410–411 (2005) 169–173.

9. J. Gubicza, N.Q. Chinh, Gy. Krállics, I. Schiller and T. Ungár, 'Microstructure of ultrafine-grained fcc metals produced by severe plastic deformation', *Current Applied Physics*, 6 (2006) 194–199.
10. J. Gubicza, N.Q. Chinh, J.L. Lábár, Z. Hegedűs and T.G. Langdon, 'Principles of self-annealing in silver processed by equal-channel angular pressing: The significance of a very low stacking fault energy', *Materials Science and Engineering*, A 527 (2010) 752–760.
11. K. Máthis, T. Krajnák, R. Kuzel and J. Gubicza, 'Structure and mechanical behaviour of interstitial-free steel processed by equal-channel angular pressing', *Journal of Alloys and Compounds*, 509 (2011) 3522–3525.
12. S.V. Dobatkin, J.A. Szpunar, A.P. Zhilyaev, J.-Y. Cho and A.A. Kuznetsov, 'Effect of the route and strain of equal-channel angular pressing on structure and properties of oxygen-free copper', *Materials Science and Engineering*, A 462 (2007) 132–138.
13. T. Ungár, G. Tichy, J. Gubicza and R.J. Hellmig, 'Correlation between subgrains and coherently scattering domains', *Powder Diffraction*, 20 (2005) 366–375.
14. J.Y. Huang, Y.T. Zhu, H. Jiang and T.C. Lowe, 'Microstructures and dislocation configurations in nanostructured Cu processed by repetitive corrugation and straightening', *Acta Materialia*, 49 (2001) 1497–1505.
15. Y.T. Zhu, J.Y. Huang, J. Gubicza, T. Ungár, Y.M. Wang, E. Ma and R.Z. Valiev, 'Nanostructures in Ti processed by severe plastic deformation', *Journal of Materials Resources*, 18 (2003) 1908–1917.
16. J. Gubicza, N.H. Nam, L. Balogh, R.J. Hellmig, V.V. Stolyarov, Y. Estrin and T. Ungár, 'Microstructure of severely deformed metals determined by X-ray peak

profile analysis', *Journal of Alloys and Compounds*, 378 (2004) 248–252.

17. O.V. Mishin, D. Juul Jensen and N. Hansen, 'Microstructures and boundary populations in materials produced by equal channel angular extrusion', *Materials Science and Engineering*, A 342 (2003) 320–328.
18. A.P. Zhilyaev, B.-K. Kim, J.A. Szpunar, M.D. Baro and T.G. Langdon, 'The microstructural characteristics of ultrafine-grained nickel', *Materials Science and Engineering*, A 391 (2005) 377–389.
19. S.S. Hazra, A.A. Gazder, A. Carman and E.V. Pereloma, 'Effect of Cold Rolling on as–ECAP Interstitial Free Steel', *Metallurgical and Materials Transactions*, A 42 (2011) 1334–1348.
20. Y.J. Chen, Y.J. Li, J.C. Walmsley, S. Dumoulin, S.S. Gireesh, S. Armada, P.C. Skareta and H.J. Roven, 'Quantitative analysis of grain refinement in titanium during equal channel angular pressing', *Scripta Materialia*, 64 (2011) 904–907.
21. H. Hallberg, M. Wallin and M. Ristinmaa, 'Modeling of continuous dynamic recrystallization in commercial-purity aluminum', *Materials Science and Engineering*, A 527 (2010) 1126–1134.
22. T. Ungár, J. Gubicza, G. Ribárik and A. Borbély, 'Crystallite size distribution and dislocation structure determined by diffraction profile analysis: principles and practical application to cubic and hexagonal crystals', *Journal of Applied Crystallography*, 34 (2001) 298–310.
23. D. Fátay, E. Bastarash, K. Nyilas, S. Dobatkin, J. Gubicza and T. Ungár, 'X-ray diffraction study on the microstructure of an Al–Mg–Sc–Zr alloy deformed by high-pressure torsion', *Zeitschrift fur Metallkunde*, 94 (2003) 842–847.

24. P.B. Escaig, 'Sur le glissement dévié des dislocations dans la structure cubique à faces centrées', *Journal de Physique*, 29 (1968) 225–239.
25. A.S. Argon and W.C. Moffatt, 'Climb of extended edge dislocations', *Acta Metallurgica*, 29 (1981) 293–299.
26. A.H.W. Ngan and M. Wen, 'Atomistic simulations of Paidar–Pope–Vitek lock formation in Ni_3Al', *Computational Materials Science*, 23 (2002) 139–145.
27. T. Ungár, E. Schafler and J. Gubicza, 'Microstructure of Bulk Nanomaterials Determined by X-Ray Line Profile Analysis', in: M.J. Zehetbauer and Y.T. Zhu (eds), *Bulk Nanostructured Materials*, WILEY–VCH, (2009) pp.361–386 (ISBN: 978–3–527–31524–6).
28. J. Gubicza, G. Ribárik, G.R. Goren-Muginstein, A.R. Rosen and T. Ungár, 'The density and the character of dislocations in cubic and hexagonal polycrystals determined by X-ray diffraction', *Materials Science and Engineering*, A309–310 (2001) 60–63.
29. T. Ungár and J. Gubicza, 'Dislocation structure and crystallite size distribution in hexagonal nanomaterials from X-ray peak profile analysis', *Zeitschrift fur Metallkunde*, 93 (2002) 694–698.
30. Y.T. Zhu, J.Y. Huang, J. Gubicza, T. Ungár, Y.M. Wang, E. Ma and R.Z. Valiev, 'Nanostructures in Ti processed by severe plastic deformation', *Journal of Materials Resources*, 18 (2003) 1908–1917.
31. P. Chatterjee and S.P. Sen Gupta, 'An X-ray diffraction study of strain localization and anisotropic dislocation contrast in nanocrystalline titanium', *Philosophical Magazine*, A 81 (2001) 49–60.
32. K. Máthis, J. Gubicza and N.H. Nam, 'Microstructure and mechanical behavior of AZ91 Mg alloy processed by equal channel angular pressing', *Journal of Alloys and Compounds*, 394 (2005) 194–199.

33. S.R. Agnew and O. Duygulu, 'A mechanistic understanding of the formability of magnesium: examining the role of temperature on the deformation mechanisms', *Materials Science Forum*, 419–422 (2003) 177–188.
34. M.H. Yoo, 'Slip, twinning, and fracture in hexagonal close-packed metals', *Metallurgical and Materials Transactions*, A 12 (1981) 409–418.
35. R. Lapovok, Y. Estrin, M.V. Popov and T.G. Langdon, 'Enhanced Superplasticity in a Magnesium Alloy Processed by Equal-Channel Angular Pressing with a Back-Pressure', *Advanced Engineering Materials*, 10 (2008) 429–433.
36. R.B. Figueiredo and T.G. Langdon, 'Principles of grain refinement and superplastic flow in magnesium alloys processed by ECAP', *Materials Science and Engineering*, A 501 (2009) 105–114.
37. R.B. Figueiredo and T.G. Langdon, 'Principles of grain refinement in magnesium alloys processed by equal-channel angular pressing', *Journal of Materials Science*, 44 (2009) 4758–4762.
38. J. Koike, 'New deformation mechanisms in fine-grain Mg alloys', *Materials Science Forum*, 419–422 (2003) 189–194.
39. A. Galiyev, R. Kaibyshev and G. Gottstein, 'Correlation of plastic deformation and dynamic recrystallization in magnesium alloy ZK60', *Acta Materialia* 49 (2001) 1199–1207.
40. J. Gubicza, S.V. Dobatkin, E. Khosravi, A.A. Kuznetsov and J.L. Lábár, 'Microstructural stability of Cu processed by different routes of severe plastic deformation', *Materials Science and Engineering*, A 528 (2011) 1828–1832.
41. E. Schafler, 'Effects of releasing the hydrostatic pressure on the nanostructure after severe plastic deformation of Cu', *Scripta Materialia*, 62 (2010) 423–426.

42. M.J. Zehetbauer, H.P. Stuwe, A. Vorhauer, E. Schafler and J. Kohout, 'The role of hydrostatic pressure in severe plastic deformation', *Advanced Engineering Materials*, 5 (2003) 330–337.
43. E. Korznikova, E. Schafler, G. Steiner and M. Zehetbauer, 'Measurements of Vacancy type Defects in SPD deformed Nickel', *Proceedings of the 4th International Symposium on Ultrafine Grained Materials 2006 TMS Annual Meeting*, TMS, Warrendale, PA, 2006, pp.97–102.
44. M. Zehetbauer, G. Steiner, E. Schafler, A. Korznikov and E. Korznikova, 'Deformation induced vacancies with severe plastic deformation: Measurements and modelling', *Materials Science Forum*, 503–504 (2006) 57–64.
45. D. Setman, E. Schafler, E. Korznikova and M.J. Zehetbauer, 'The presence and nature of vacancy type defects in nanometals detained by severe plastic deformation', *Materials Science and Engineering*, A 493 (2008) 116–122.
46. A.R. Kilmametov, G. Vaughan, A.R. Yavari, A. LeMoulec, W.J. Botta and R.Z. Valiev, 'Microstructure evolution in copper under severe plastic deformation detected by in situ X-ray diffraction using monochromatic synchrotron light', *Materials Science and Engineering*, A 503 (2009) 10–13.
47. D. Setman, M.B. Kerber, E. Schafler and M.J. Zehetbauer, 'Activation Enthalpies of Deformation-Induced Lattice Defects in Severe Plastic Deformation Nanometals Measured by Differential Scanning Calorimetry', *Metallurgical and Materials Transactions*, A 41 (2010) 810–815.
48. J.P. Hirth and J. Lothe, *Theory of Dislocations*, John Wiley, New York, 1982.
49. L.E. Murr, *Interfacial Phenomena in Metals and Alloys*, Addison Wesley, Reading, MA, 1975.

50. M. Furukawa, Z. Horita, M. Nemoto and T.G. Langdon, 'Review: processing of metals by equal-channel angular pressing', *Journal of Materials Science*, 36 (2001) 2835–2843.
51. Y. Iwahashi, Z. Horita, M. Nemoto and T.G. Langdon, 'The shearing characteristics associated with equal-channel angular pressing', *Acta Materialia*, 46 (1998) 3317–3331.
52. L. Balogh, T. Ungár, Y. Zhao, Y.T. Zhu, Z. Horita, C. Xu and T.G. Langdon, 'Influence of stacking-fault energy on microstructural characteristics of ultrafine-grain copper and copper-zinc alloys', *Acta Materialia*, 56 (2008) 809–820.
53. A.P. Zhilyaev, K. Oh-ishi, T.G. Langdon and T.R. McNelley, 'Microstructural evolution in commercial purity aluminum during high-pressure torsion', *Materials Science and Engineering*, A 410–411 (2005) 277–280.
54. N. Rangaraju, T. Raghuram, B. Vamsi Krishna, K. Prasad Rao and P. Venugopal, 'Effect of cryo-rolling and annealing on microstructure and properties of commercially pure aluminium', *Materials Science and Engineering*, A 398 (2005) 246–251.
55. J. Gubicza, N.Q. Chinh, P. Szommer, A. Vinogradov and T.G. Langdon, 'Microstructural characteristics of pure gold processed by equal-channel angular pressing', *Scripta Materialia*, 56 (2007) 947–950.
56. A.P. Zhilyaev, J. Gubicza, G. Nurislamova, Á. Révész, S. Suriñach, M.D. Baró and T. Ungár, 'Microstructural characterization of ultrafine-grained nickel', *Physica Status Solidi*, (a) 198 (2003) 263–271.
57. P.V. Liddicoat, X.-Z. Liao, Y. Zhao, Y. Zhu, M.Y. Murashkin, E.J. Lavernia, R.Z. Valiev and S.P. Ringer, 'Nanostructural hierarchy increases the strength of

aluminium alloys', *Nature Communications*, 1 (2010) Art. No. 63.

58. S. Qu, X.H. An, H.J. Yang, C.X. Huang, G. Yang, Q.S. Zang, Z.G. Wang, S.D. Wua and Z.F. Zhang, 'Microstructural evolution and mechanical properties of Cu–Al alloys subjected to equal channel angular pressing', *Acta Materialia*, 57 (2009) 1586–1601.
59. J. Gubicza, I. Schiller, N.Q. Chinh, J. Illy, Z. Horita and T.G. Langdon, 'The effect of severe plastic deformation on precipitation in supersaturated Al-Zn-Mg alloys', *Materials Science and Engineering*, A 460–461 (2007) 77–85.
60. M. Dopita, M. Janecek, R. Kuzel, H.J. Seifert and S. Dobatkin, 'Microstructure evolution of CuZr polycrystals processed by high-pressure torsion', *Journal of Materials Science*, 45 (2010) 4631–4644.
61. R. Kuzel, M. Janecek, Z. Matej, J. Cizek, M. Dopita and O. Srba, 'Microstructure of Equal-Channel Angular Pressed Cu and Cu-Zr Samples Studied by Different Methods', *Metallurgical and Materials Transactions*, A 41 (2010) 1174–1190.
62. H.J. Frost and M.F. Ashby, *Deformation-Mechanism Maps: The Plasticity and Creep of Metals and Ceramics*, Oxford, Pergamon, 1982.
63. S. Schumacher, R. Birringer, R. Straus and H. Gleiter, 'Diffusion of silver in nanocrystalline copper between 303 and 373 K', *Acta Metallurgica*, 37 (1989) 2485–2488.
64. Y.S. Li, Y. Zhang, N.R. Tao and K. Lu, 'Effect of the Zener-Hollomon parameter on the microstructures and mechanical properties of Cu subjected to plastic deformation', *Acta Materialia*, 57 (2009) 761–772.
65. A. Mishra, B.K. Kad, F. Gregori and M.A. Meyers, 'Microstructural evolution in copper subjected to severe

plastic deformation: Experiments and analysis', *Acta Materialia*, 55 (2007) 13–28.

66. A. Dubravina, M.J. Zehetbauer, E. Schafler and I.V. Alexandrov, 'Correlation between domain size obtained by X-ray Bragg profile analysis and macroscopic flow stress in severely plastically deformed copper', *Materials Science and Engineering*, A 387–389 (2004) 817–821.
67. R. Kuzel, Z. Matej, V. Cherkaska, J. Pesicka, J. Cizek, I. Prochazka and R.K. Islamgaliev, 'Microstructure of severely deformed metals and its thermal stability', *Journal of Alloys and Compounds*, 378 (2004) 242–247.
68. E. Schafler, A. Dubravina, B. Mingler, H.P. Karnthaler and M. Zehetbauer, 'On the microstructure of HPT processed Cu under variation of deformation parameters', *Materials Science Forum*, 503–504 (2006) 51–56.
69. J. Gubicza, N.Q. Chinh, J.L. Lábár, Z. Hegedűs and T.G. Langdon, 'Twinning and dislocation activity in silver processed by severe plastic deformation', *Journal of Materials Science*, 44 (2009) 1656–1660.
70. M. Kawasaki, Z. Horita and T.G. Langdon, 'Microstructural evolution in high purity aluminum processed by ECAP', *Materials Science and Engineering*, A 524 (2009) 143–150.
71. C. Xu, Z. Horita and T.G. Langdon, 'Microstructural evolution in an aluminum solid solution alloy processed by ECAP', *Materials Science and Engineering*, A 528 (2011) 6059–6065.
72. X.Z. Liao, A.R. Kilmametov, R.Z. Valiev, H. Gao, X. Li, A.K. Mukherjee, J.F. Bingert and Y.T. Zhu, 'High-pressure torsion-induced grain growth in electrodeposited nanocrystalline Ni', *Applied Physics Letters*, 88 (2006) 021909.

73. J. Cizek, M. Janecek, O. Srba, R. Kuzel, Z. Barnovska, I. Prochazka and S. Dobatkin, 'Evolution of defects in copper deformed by high-pressure torsion', *Acta Materialia*, 59 (2011) 2322–2329.
74. B.R. Watts, 'Conduction Electrons Scattering in Dislocated Metals' in: F.R.N. Nabarro (ed.), *Dislocation in Solids*, vol. 8, Elsevier, Amsterdam, 1989, pp.175–419.
75. H.J. Wollenberger, 'Point Defects' in: R.W. Cahn and P. Haasen (eds), *Physical Metallurgy*, vol. 9, Elsevier, Amsterdam, 1983, pp.1189–1221.
76. P. Varotsos and K. Eftaxias, 'High-temperature vacancy concentration in Cu', *Physical Review*, B 40 (1989) 9963–9964.
77. T. Ungar, E. Schafler, P. Hanak, S. Bernstorff and M. Zehetbauer, 'Vacancy production during plastic deformation in copper determined by in situ X-ray diffraction', *Materials Science and Engineering*, A 462 (2007) 398–401.

3

Defect structure in low stacking fault energy nanomaterials

Abstract: In this chapter, the effect of low stacking fault energy on defect structure in nanomaterials is summarised. The low value of stacking fault energy leads to a large degree of dislocation dissociation into partials, which strongly hinders the cross-slip and climb of dislocations. As a consequence, a relatively large dislocation density develops during severe plastic deformation of low stacking fault energy metallic materials. Additionally, the low stacking fault energy is accompanied by a small value of twin boundary energy, resulting in a significant twinning activity during plastic deformation that alters grain refinement mechanisms. It is also shown that the reduction of grain size increases the splitting distance between partials.

Key words: low stacking fault energy, dislocation dissociation, twinning, untwinning.

3.1 Effect of low stacking fault energy on cross-slip and climb of dislocations

The rearrangement and annihilation of screw and edge dislocations during plastic deformation occur by cross-slip and climb, respectively. The probability of the occurrence of these mechanisms is dependent upon the degree of dislocation dissociation, therefore the low stacking fault energy (SFE) considerably affects the evolution of the dislocation structure during SPD-processing of ultrafine-grained (UFG) or nanomaterials. In face-centred cubic (fcc) metals the lattice dislocations are dissociated into Shockley partials. As an example, the high-resolution transmission electron microscopy (HRTEM) image in Figure 3.1 shows a dissociated screw dislocation in Ag processed by eight passes of equal-channel angular pressing (ECAP) at room temperature (RT) with a mean grain size of 200 nm. In this case, the lattice screw dislocation splits into two partials that can be described by the following reaction of Burgers vectors [1, 2]:

$$\frac{1}{2}\left[01\bar{1}\right] \rightarrow \frac{1}{6}\left[11\bar{2}\right] + \frac{1}{6}\left[\bar{1}2\bar{1}\right]$$

Figure 3.1 HRTEM image showing a dissociated screw dislocation bounded by partials in Ag processed by eight ECAP passes (right) and a picture illustrating the corresponding crystallographic directions and lattice planes (left)

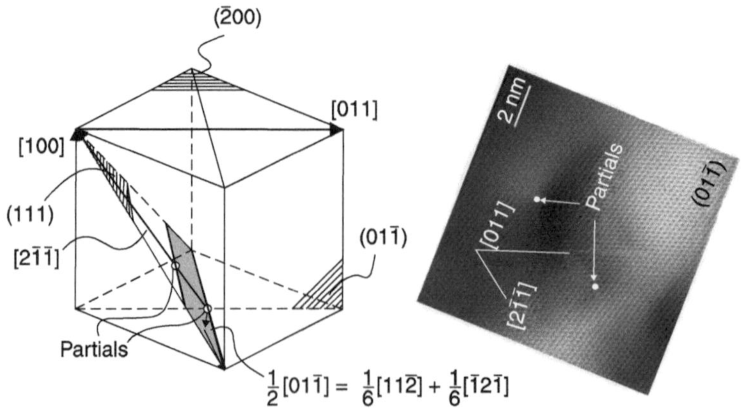

The equilibrium splitting distance between Shockley partial dislocations, d_p, may be determined by equating the repulsive force originating from the stress fields of the partials (f_r) to the attractive force resulting from the formation of a stacking fault between the two partials (f_a). The former and the latter forces per unit dislocation length can be expressed as [1]:

$$f_r = \frac{Gb^2(2-\nu)}{24\pi d_p(1-\nu)}\left[1-\frac{2\nu}{2-\nu}\cos 2\Phi\right], \tag{3.1}$$

and

$$f_a = \gamma, \tag{3.2}$$

respectively, where G is the shear modulus, b is the magnitude of Burgers vector of full lattice dislocation, ν is the Poisson's ratio, γ is the SFE and Φ is the angle between the Burgers vector and the line vector of lattice dislocation (Φ = 0° and 90° for screw and edge dislocations, respectively). Then the equilibrium splitting distance is expressed as [1]:

$$d_p = A\frac{Gb^2}{\gamma}, \tag{3.3}$$

where A is a constant having values of ~0.019 and ~0.048 for screw and edge dislocations, respectively [1]. The values of the equilibrium splitting distances for screw and edge dislocations calculated using Equation 3.3 are listed in Table 3.1. The formula for the equilibrium splitting distance that takes the elastic anisotropy into account can be found in Hirth and Lothe (1982) [2]. It is noted that the splitting distance between partials for the screw dislocation shown in the HRTEM image of Figure 3.1 is about 5 nm – two times larger than the equilibrium value in Ag – which is most probably caused by internal stresses. The shear stresses acting on the glide plane of dissociated dislocations may increase or decrease the splitting distance between partials.

Table 3.1 **The equilibrium splitting distance (d_p) for screw and edge dislocations, the constriction energy (W_0) and the waiting time for cross-slip of screw dislocations (t_{cs}) in various pure fcc metals**

Material	γ [mJ/m2]	d_p (screw) [nm]	d_p (edge) [nm]	W_0 [eV]	t_{cs} [s]
Pt	322	0.28	0.70	0.24	8×10^{-10}
Pd	180	0.38	0.91	0.28	5×10^{-9}
Al	166	0.24	0.61	0.09	3×10^{-12}
Ni	125	0.77	1.95	1.02	4×10^{5}
Cu	62	0.94	2.38	0.78	3
Au	38	1.12	2.83	0.68	3×10^{-2}
Ag	19	2.51	6.33	1.95	4×10^{20}

Note: The values of the stacking fault energy (γ) are taken from Hirth and Lothe (1982) and Murr (1975) [2, 3].

Figure 3.2 shows the forces acting on partials of a dissociated screw dislocation for four different shear stress directions. The direction of a force is determined by the component of the partial Burgers vector lying parallel to the stress as it follows from the general formula of the force acting on a dislocation: $f = (\sigma\, \boldsymbol{b}) \times \boldsymbol{l}$, where σ is the stress tenor, $\boldsymbol{b}$ is the Burgers vector and $\boldsymbol{l}$ is the line vector of the dislocation. Figures 3.2a and b show that if the stresses are parallel to the line vectors of partials (or to the Burgers vector of the undissociated screw dislocation), the forces yield only slip of the dissociated screw dislocation without any change of the splitting distance since the partial Burgers vectors have the same components parallel to the stress direction. However, if the stress is perpendicular to the line vectors of partials (or to the Burgers vector of the undissociated screw dislocation), the induced forces cause a decrease (Figure 3.2c) or increase

Figure 3.2 **The forces f_1 and f_2 acting on partials in dissociated screw dislocations due to the shear stress, τ, in the glide plane. The glide plane is lying in the sheet and the stacking fault between partials is marked by a light grey colour. The Burgers vectors of partials are denoted by b_1 and b_2. The components of these vectors lying parallel and perpendicular to the dislocation line vector, l, are also presented**

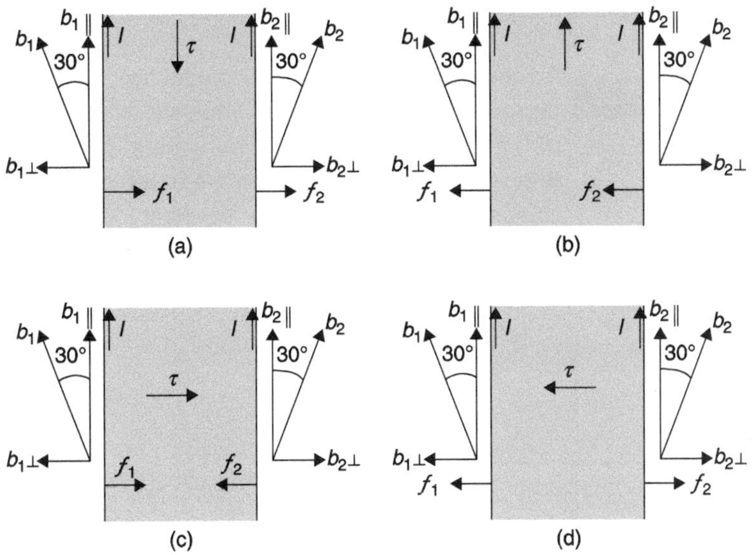

(Figure 3.2d) of the splitting distance due to the different orientations of the partial Burgers vectors' components parallel to the stress direction. In the case of a dissociated edge dislocation, the shear stresses acting parallel to the Burgers vector of the undissociated dislocation (perpendicular to the line vector) yield only slip without changing the splitting distance while the stresses perpendicular to the Burgers vector of the undissociated dislocation (parallel to the line vector) result in change of the splitting distance.

It is important to note that the equilibrium splitting distance between partials depends strongly on the grain size when the grain diameter falls under a few tens of nanometres. For modelling this effect, a grain with a square (111) slip plane and a size *d* is considered as depicted in Figure 3.3 [4]. Under an external stress, a leading Shockley partial dislocation, denoted by XEFY in Figure 3.3, is emitted from the grain boundary XY. The Burgers vector of the leading

Figure 3.3 A schematic illustration of a model for dissociated dislocation with wide stacking fault (SF) ribbon in a rectangular nanograin. Two Shockley partial dislocations are emitted consecutively from the grain boundary XY, with their ends pinned at X and Y [4]

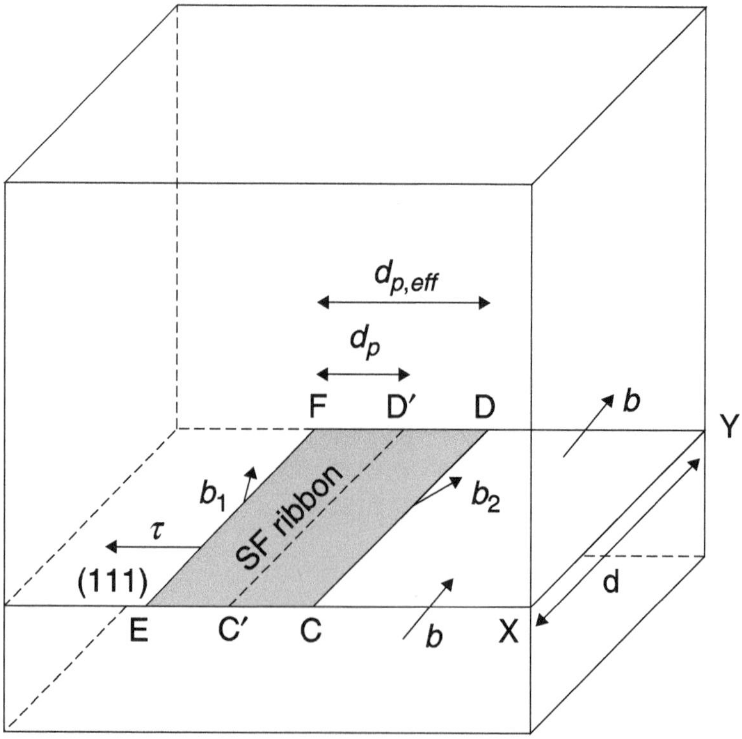

partial is $b_1 = \frac{a}{6}\left[2\bar{1}\bar{1}\right]$. The ends of this partial dislocation are pinned at X and Y, so that two segments of partial dislocation lines XE and FY are deposited on grain boundaries. Similarly, a trailing Shockley partial dislocation, denoted by XCDY, with the Burgers vector $b_2 = \frac{a}{6}\left[1\bar{2}1\right]$ is emitted from the grain boundary XY. Partial dislocation lines EF and CD are separated by a stacking fault ribbon, and they would form a screw dislocation with the Burgers vector $b = \frac{a}{2}\left[1\bar{1}0\right]$, if collapsed together. The two partial dislocations react to form two full edge dislocation segments, XC and DY, with the Burgers vector $b = \frac{a}{2}\left[1\bar{1}0\right]$ at the grain boundaries [4].

Without the full dislocation segments XC and DY, the equilibrium splitting distance between the partials, d_p, would be determined by equating the forces given by Equations 3.1 and 3.2. In this case, the stacking fault ribbon would lie between the segment EF and the dashed line C′D′ in Figure 3.3. However, assuming that the leading partial dislocation line EF stays stationary, the existence of dislocation segments XC and DY shifts the trailing partial dislocation line C′D′ to the position CD. This motion causes a reduction of the total energy of the whole dislocation configuration since the full dislocation segments CC′ and D′D are replaced by partial dislocations having smaller energies due to their smaller Burgers vector. This decrease of energy overwhelms the higher energy of the larger stacking fault ribbon. Thus, the effective splitting distance between partials, $d_{p,eff}$, for the grain size d can be determined by equating the attractive and repulsive forces as:

$$\gamma \cdot d = \frac{Gb^2(2-3\nu)}{24\pi \cdot d_{p,eff} \cdot (1-\nu)} \cdot d + 2 \cdot \left(E_f - E_p\right), \qquad (3.4)$$

where the term in the left-hand side gives the attractive force due to the stacking fault ribbon, the first term in the right side accounts for the repulsive force between the partials and the second term is a repulsive force resulting in the energy gain when the full dislocation segments CC′ and D′D are replaced by partial dislocations. E_f and E_p are the energies of full and partial dislocations per unit length given as [4]:

$$E_f = \frac{Gb^2}{4\pi(1-\nu)} \ln\left(\frac{d}{b}\right) \tag{3.5}$$

and

$$E_p = \frac{Gb_p^2(1-0.25\cdot\nu)}{4\pi(1-\nu)} \ln\left(\frac{d}{b}\right), \tag{3.6}$$

respectively, where b_p is the magnitude of partial Burgers vector. Applying the relations $b = \frac{a}{\sqrt{2}}$ and $b_p = \frac{a}{\sqrt{6}}$, where a is the lattice parameter for fcc structure, and using Equation 3.3 valid for UFG or coarse-grained materials, $d_{p,eff}$ can be given as [4]:

$$d_{p,eff} = \frac{\gamma}{\gamma - \frac{G\cdot a^2(8+\nu)}{48\pi\cdot(1-\nu)\cdot d} \ln\left(\frac{\sqrt{2}d}{a}\right)} \cdot d_p. \tag{3.7}$$

Taking into account a stress, τ, acting perpendicular to the partial segments CD and EF (see Figure 3.3), Equation 3.7 is modified as [4]:

$$d_{p,eff} = \frac{\gamma}{\gamma - \frac{G\cdot a^2(8+\nu)}{48\pi\cdot(1-\nu)\cdot d} \ln\left(\frac{\sqrt{2}d}{a}\right) - \frac{\tau\cdot a}{\sqrt{6}}} \cdot d_p \tag{3.8}$$

It is emphasised that the previous consideration can be applied only for the case when a single dislocation is emitted from the grain boundary and the width of the source equals the grain size. Therefore, Equations 3.7 and 3.8 are valid for very small grain sizes where dislocation sources inside grains do not operate, i.e. below ~20 nm. Figure 3.4 illustrates the variation of the effective equilibrium splitting distance between partials in nanostructured Al as a function of grain size according to Equation 3.7 where G = 26 GPa, a = 0.405 nm, ν = 0.34 and γ = 166 mJ/m^2 are used. For the grain size of ~20 nm, $d_{p,eff}$ is about two times larger than the value of 0.24 nm obtained for coarse-grained Al (see Table 3.1). The splitting distance increases steeply with further reduction of the grain size and at 7.17 nm it reaches the grain diameter. Under this critical grain size, Equation 3.7 cannot be applied for calculation of the splitting distance as the stacking fault spreads across the grain and the splitting distance equals the grain size. The critical grain size values for Ni, Cu and Ag

Figure 3.4 The effective equilibrium splitting distance between partials, $d_{p,eff}$ in nanocrystalline Al as a function of grain size (solid line). The dashed line corresponds to the equilibrium splitting distance characteristic for coarse-grained counterparts

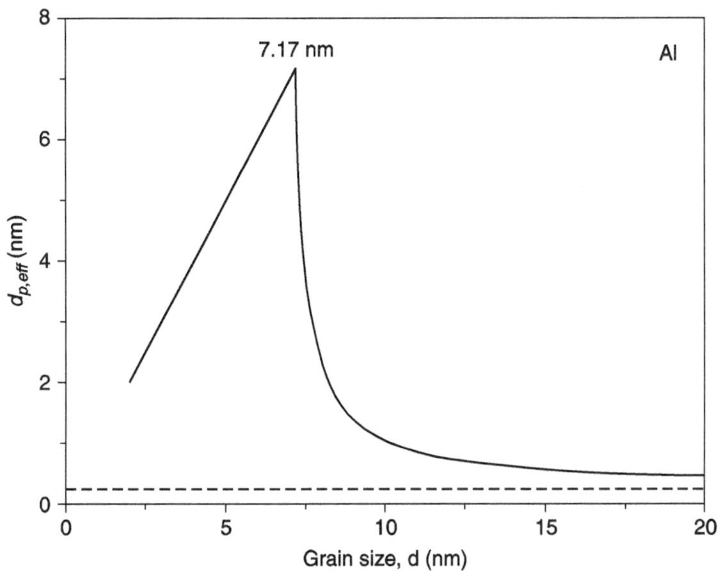

having smaller SFE than that for Al, are 36.3, 43.5 and 140 nm, respectively. These critical values seem to be too large for applying the model depicted in Figure 3.3; therefore, Equation 3.7 cannot be used for medium or low SFE nanomaterials. In these materials, below ~20 nm, the plasticity in the grain interiors is controlled rather by partial dislocations than dissociated lattice dislocations. HRTEM experiments support the existence of wide stacking fault ribbons in Al nanomaterials [5]. It is noted that the actual stresses acting on the partial dislocation segments may alter the splitting distance very much, especially in nanomaterials where these stresses are usually very high.

The equilibrium splitting distance is exceptionally high in UFG Ag among the pure fcc metals because of the very low SFE (see Table 3.1). This high degree of dislocation dissociation impedes motion by cross-slip. In the cross-slip model developed by Friedel [6], cross-slip occurs in fcc crystals by the formation of an initial constriction in the dissociated dislocation and the subsequent extension of this constricted segment onto the cross-slip plane as depicted in Figure 3.5. The cross-slip process is stress-assisted because the activation energy of the process is reduced by the shear stress, σ', pushing the partials towards each other on the initial glide plane and by the shear stress, σ_s, pulling the partials in opposite directions on the cross-slip plane (see Figure 3.5). The energy, W, required for cross-slip is given by the relationship [7]:

$$W = W_0\left(1 - 1.2\frac{\sigma_s b}{\gamma} - 1.5\frac{\sigma' b}{\gamma}\right), \tag{3.9}$$

where W_0 is the constriction energy, which is given by:

$$W_0 = \frac{Gb^2 d_p}{37}\left(\ln\left(\frac{2\sqrt{3}d_p}{b}\right)\right)^{1/2}. \tag{3.10}$$

Figure 3.5 **Cross-slip of a dissociated screw dislocation. σ' and σ_s are the shear stresses pushing the partials towards each other on the initial glide plane (S_1) and pulling the partials in opposite directions on the cross-slip plane (S_2), respectively**

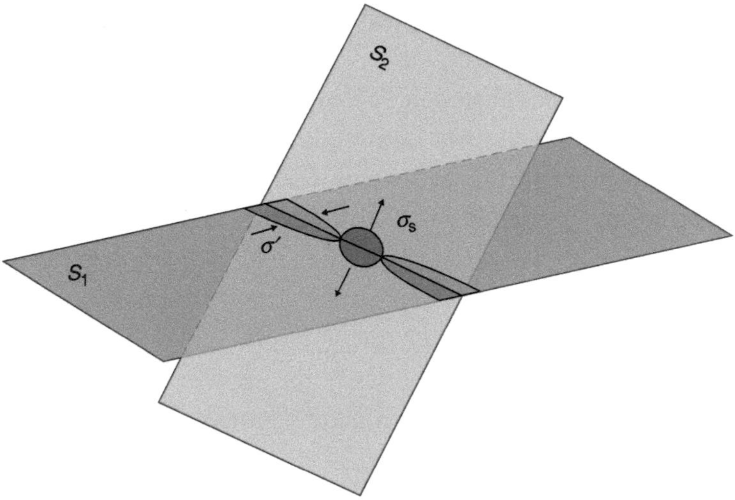

Although there are other slightly different relationships for the constriction energy [8, 9], the order of magnitude of W_0 remains similar and all relationships predict an increasing constriction energy with increasing d_p/b.

The energy required for cross-slip is supplied by the thermal fluctuations occurring in the crystalline lattice. From the thermal activation of cross-slip, it follows that the waiting time for this process, t_{cs}, may be expressed as:

$$t_{cs} = \frac{1}{\nu}\exp\left(\frac{W}{kT}\right), \tag{3.11}$$

where k is Boltzmann's constant, T is the absolute temperature and ν is the frequency for the crystal vibrations (~10^{13} Hz), which can be given as:

$$\nu = \frac{k\Theta_D}{h}, \tag{3.12}$$

where h is the Planck's constant and Θ_D is the Debye temperature listed in Baria (2004) [10]. Thus, higher values of W correspond to longer waiting times for cross-slip. The constriction energy and the waiting times for cross-slip at RT were calculated from Equations 3.9–3.12 for seven different pure fcc metals and listed in Table 3.1. Since the values of the stresses σ' and σ_s are not known, a first approximation was adopted whereby the cross-slip energy was taken as the constriction energy, W_0. Thus, the smallest and largest values of t_{cs} are related to the lowest and the highest degrees of dissociation in Al and Ag, respectively, and for metals having intermediate values of d_p (Pt, Pd, Ni, Cu and Au), the estimated values of t_{cs} lie between the extreme values for Al and Ag. The relatively high value of t_{cs} for Ni is caused by its large shear modulus (82 GPa).

For Al, Pt, Pd, Cu and Au the waiting times for cross-slip are significantly smaller than the usual total time for any SPD-processing (several minutes) so that the annihilation of the screw dislocations by cross-slip takes place during the processing operation. By contrast, the waiting time for cross-slip in Ag is of the order of $\sim 10^{20}$ s, equivalent to a waiting period of $\sim 10^{13}$ years, which means in practice that the dislocations formed during SPD are not annihilated by cross-slip without the benefit of some assisting stresses. However, there is evidence for internal stresses in the microstructures produced by SPD due to the presence of other dislocations and in practice these internal stresses may either assist or hinder the cross-slip process. For example, in Figure 3.1 the splitting distance is about twice the equilibrium value that is most probably caused by internal stresses. The activation energy for cross-slip is very sensitive to the presence of these stresses. For instance, taking one-tenth of the flow stress for Ag processed by eight ECAP passes (30 MPa) as the assisting stress of σ' and zero for σ_s, the waiting time for cross-slip is reduced to 100 s in Ag. This means that during SPD-processing of low SFE nanomaterials cross-slip can occur due to assisting stresses. In severely deformed polycrystals, the magnitude and direction of internal stresses acting on the glide planes change from grain to grain due to the distributions of the dislocations and other defects, such as twins, so that the probability of cross-slip depends upon their specific locations within the specimen. As there are locations where the stresses do not assist or even hinder cross-slip, the total dislocation density is usually very high in SPD-processed low SFE nanomaterials.

In addition to the cross-slip of screw dislocations, recovery may occur through the climb of edge dislocations. The climb velocity of an extended dislocation is [11]:

$$v_c = \frac{1050\sigma \Omega D}{bkT} c_j \left(\frac{d_p}{b} \right)^{-2}, \tag{3.13}$$

where σ is the stress acting across the extra half plane of the dislocation producing a climb force σb, Ω is the atomic volume, D is the diffusion coefficient and c_j is the number of jogs along the dislocation line. The characteristic waiting time for the annihilation of edge dislocations by climb, t_{climb}, may be approximated as the ratio of the mean spacing between dislocations ($\rho^{-1/2}$ where ρ is the average dislocation density) and the climb velocity (v_c) so that:

$$t_{\text{climb}} = \frac{\rho^{-1/2}}{v_c}. \tag{3.14}$$

It can be established from Equation 3.13 that the velocity of climb decreases with an increase in the splitting distance between partials, d_p/b, and therefore dislocation dissociation also hinders annihilation by climb. This effect also contributes to the very high dislocation density in SPD-processed low SFE UFG or nanomaterials. Equations 3.13 and 3.14 show that the waiting time for climb depends quadratically on d_p/b, but for cross-slip the waiting time depends exponentially on the splitting distance, which means that cross-slip is more sensitive to an increase in the splitting distance. There is a similar difference in the stress sensitivity of the two mechanisms. The characteristic waiting time for climb in Ag may be estimated by substituting appropriate values into Equations 3.13 and 3.14: $\sigma = 30$ MPa, which is similar to the value used for cross-slip, $\Omega/b = 10^{-19}$ m^2, $c_j = 1$, $kT = 4 \times 10^{-21}$ J, $d_p/b = 22$ from Equation 3.3 for edge dislocations and taking $\rho = 46 \times 10^{14}$ m^{-2} as the value after eight ECAP passes [12] so that $\rho^{-1/2} = 15$ nm. The value of the diffusion coefficient depends on the migration path of vacancies since D is ~10^{-35} m^2 s^{-1} [13] in bulk Ag and ~10^{-20} m^2 s^{-1} along grain boundaries and dislocations [14]. Due to the high dislocation density and small grain size, it is assumed that vacancy diffusion occurs predominantly along dislocations and grain boundaries so that D ≈ 10^{-20} m^2 s^{-1} and, from Equations 3.13 and 3.14 $t_{climb} \approx 100$ s is obtained, which is similar to that for cross-slip. However, for stress values smaller than 30 MPa, the waiting time is much higher in the case of cross-slip than climb.

The retarding effect of high degree of dislocation dissociation on cross-slip and climb results in a very large dislocation density in ultrafine-grained Ag compared with other pure fcc metals having medium or high SFE. The saturation values of the dislocation density in pure nanostructured fcc metals having various SFEs are compared in Chapter 2.

3.2 Defect structure developed in SPD-processed low stacking fault energy pure Ag

The evolution of the defect structure during ECAP-processing at RT was investigated in 99.995% (4N5) purity Ag [12]. After one pass the grain size was reduced to ~20–30 μm. The interior of the grains contains both dislocations and twins as illustrated in the TEM image of Figure 3.6a. The minimum grain size, ~200 nm, was achieved after four passes, as shown in Figure 3.6b, and this size remained essentially unchanged within experimental error after eight and 16 passes (see Figures 3.6c and 3.6d). The white arrows indicate the locations of some twin boundaries in Figure 3.6. The evolution of the dislocation density and the twin boundary frequency as determined by x-ray line profile analysis is plotted in Figure 3.7. The twin-boundary frequency is defined as the relative fraction of twin boundaries among the {111} lattice planes. After one pass of ECAP the dislocation density and the twin boundary frequency are $16 \pm 2 \times 10^{14}$ m^{-2} and 0.1 ± 0.1%, respectively. After four passes the dislocation density and the twin boundary frequency increase to $37 \pm 4 \times 10^{14}$ m^{-2} and 0.7 ± 0.1%, respectively. However, whereas the dislocation density saturates after eight passes at $46 \pm 5 \times 10^{14}$ m^{-2} and then decreases to $25 \pm 3 \times 10^{14}$ m^{-2} after 16 passes, the measurements show the twin boundary frequency increases to 0.9 ± 0.1% after eight passes and then further increases to 1.5 ± 0.1% after 16 passes [12].

The low SFE is accompanied by low twin boundary energy in Ag, which explains the observed high values of twin boundary frequency. Figure 3.8 gives the saturation twin boundary frequency achieved by ECAP at RT as a function of twin fault energy for pure fcc metals. It can be seen that

Figure 3.6 TEM images taken after (a) 1, (b) 4, (c) 8 and (d) 16 passes of ECAP: examples of twin boundaries are indicated by white arrows

Source: Reprinted from Gubicza et al. (2010) [12] with permission from Elsevier.

there is a sharp increase in the saturation twin boundary frequency with decreasing twin fault energy [15].

As it is revealed, there is a very significant difference between the observed trends for the dislocation density and the twin boundary frequency in 4N5 purity UFG Ag processed by ECAP. These differing evolutions of the dislocation and twin densities with increasing strain are consistent with the theoretical predictions of a model developed earlier by Müllner and Solenthaler [16] for the plastic straining of fcc metals and alloys having low SFE. At

Figure 3.7 The dislocation density and the twin boundary frequency as a function of number of ECAP passes for 4N5 purity Ag

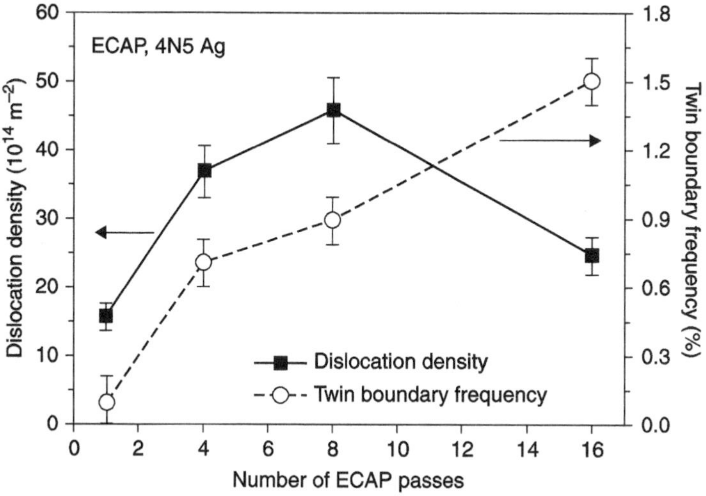

Source: Reprinted from Ref. [12] with permission from Elsevier.

Figure 3.8 The saturation twin boundary frequency achieved by ECAP at RT versus the twin fault energy, γ_T for pure fcc metals

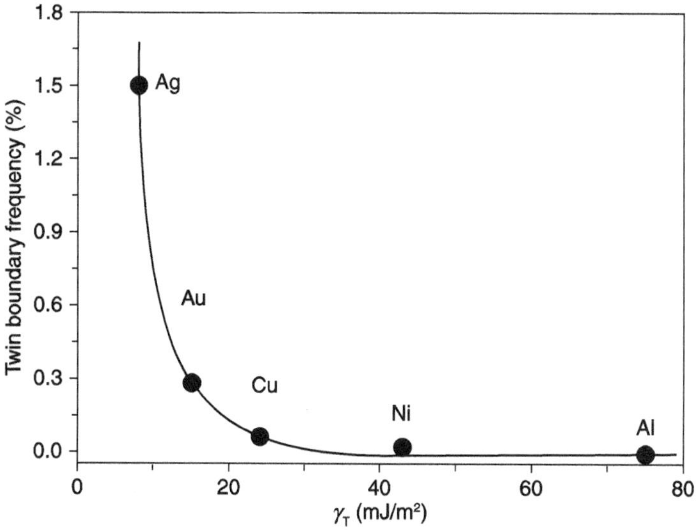

relatively low strains, the deformation behaviour in these materials is controlled mainly by the planar glide of dissociated dislocations and the dislocation density increases rapidly. As the deformation proceeds, dislocation pile-ups form at glide obstacles such as Lomer-Cottrell locks and grain boundaries. In fcc metals having very low SFE, these glide obstacles strongly hinder the further activity of lattice dislocations because the non-conservative motion of dislocations is difficult due to the high degree of dissociation. At these obstacles, if the local stress exceeds the critical stress required for twin nucleation, plasticity continues by twinning. With increasing strain as in ECAP, the dislocation density further increases, together with the locations of these high stress concentrations, leading to an increased density of twins. However, these twins also obstruct the glide of lattice dislocations, thereby hindering the operation of dislocation sources and leading to a reduction in the rate of dislocation production.

Figure 3.7 shows that there is a well-defined saturation in the dislocation density and thereafter, at higher strains, the dislocation density decreases while the twin probability frequency continues to increase. This reduction in the dislocation density at high strains may be explained in terms of the nature of the twinning mechanism. Several possible mechanisms have been proposed for twin formation including (i) by a pole mechanism [17], (ii) by the dissociation of lattice dislocations into Schockley and Frank partials at Lomer-Cottrell locks [18], (iii) by the nucleation of three-layer twins by dissociation of co-planar lattice dislocations into Schockley partials [19] and/or (iv) by the emission of twinning partials from grain boundaries [20, 21]. In the fourth process, it may be assumed that the partials form by a dissociation of non-geometrically necessary lattice dislocations in the grain boundaries [20]. Figures 3.9a and 3.9b illustrate the formation

Figure 3.9 The formation of twins at Lomer-Cottrell locks (a) and grain boundaries (b) by dissociation of lattice dislocations into twinning partials and their slip on successive {111} planes. The Schockley-partials are denoted by 'L'

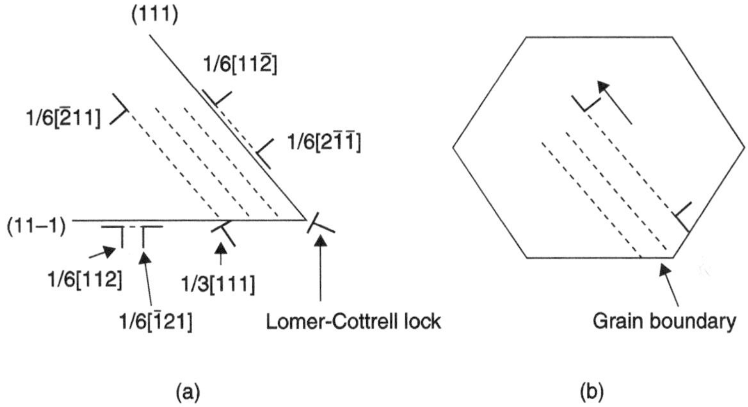

of twins at Lomer-Cottrell locks and grain boundaries, respectively. In the former case, the reaction can be given in terms of the Burgers vectors as:

$$\frac{1}{6}[112]+\frac{1}{6}[\bar{1}21]\rightarrow\frac{1}{3}[111]+\frac{1}{6}[\bar{2}11]$$

All of the mechanisms of deformation twinning are based inherently on the dissociation of lattice dislocations into twinning partials and their slip on successive {111} planes. Therefore, if these partials then move to the grain boundaries, the formation of twins will contribute to the annihilation of lattice dislocations. As a consequence, the decrease in the dislocation density between 8 and 16 passes, as shown in Figure 3.7, is associated with an increase in the twin boundary frequency. It is important to note that the dislocation sources are incapable of compensating for the dislocations that have disappeared by twinning because the operation of the sources is also obstructed by the newly nucleated twins. Untwinning may also occur during the deformation process due to the interaction between the dislocations and twins [16], but the present increase in the twin probability with increasing strain suggests that the twin production rate is larger than that for untwinning. The untwinning process is depicted in Figure 3.10. The notations of dislocations in Figure 3.10 are explained with the help of the Thompson tetrahedron

Figure 3.10 The untwinning process. The twin boundaries are indicated by 'TB'. The partial dislocations are denoted as is usual in a standard Thompson-tetrahedron (see Figure 3.11). When a dissociated dislocation meets a twin lamella, the leading partial (αB) splits into a sessile star-rod dislocation ($\alpha\delta$) and a glissile Shockley-partial (δB), which slips on a layer of the twin lamella resulting in untwinning on that layer (a). If dislocations on successive slip planes meet twin lamella, a complete disappearance of twin segment can occur (b) [16]

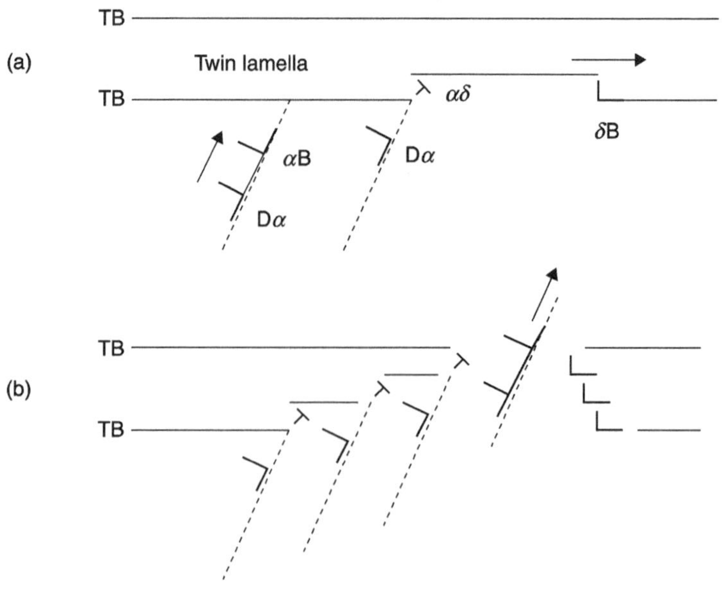

presented in Figure 3.11. The slip planes and the Burgers vectors of lattice dislocations are represented by the faces and the edges of the tetrahedron. The Burgers vectors of Schockley partial dislocations connect the vertices of the tetrahedron with the centres of the faces. For instance, the origin and the endpoint of the Burgers vector of the Schockley partial dislocation Dα in Figure 3.10 are the points D and α in Figure 3.11. When the leading partial αB of an initial dissociated dislocation in Figure 3.10a meets a barrier twin boundary, it may split into a sessile stair-rod dislocation $\alpha\delta$ and a glissile Schockley-partial δB due to appropriate stresses. The Burgers vectors of stair rod dislocations connect the

Figure 3.11 **(a) Thompson tetrahedron ABCD and (b) its two-dimensional representation illustrating the possible slip planes and the Burgers vectors of dislocations in an fcc crystal. The four faces of the tetrahedron corresponding to the slip planes are denoted by *a*, *b*, *c* and *d* while the centres of the faces are indicated as α, β, γ and δ**

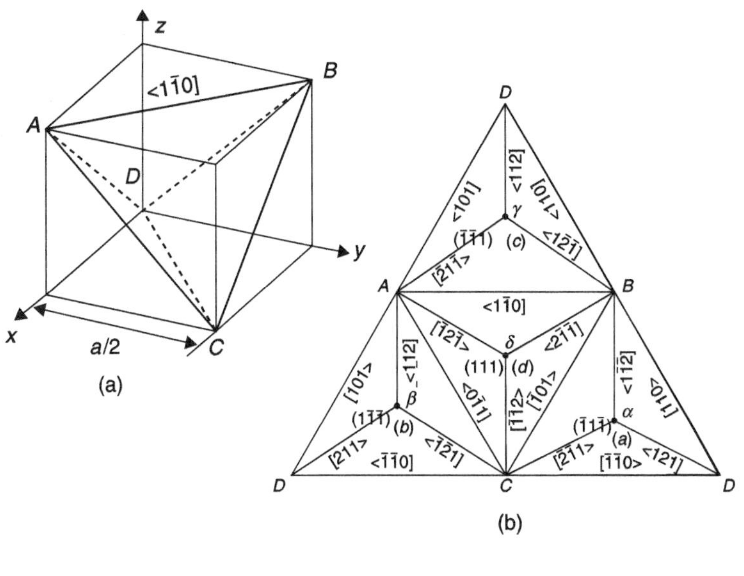

centres of the faces of Thompson tetrahedron. For example, the stair-rod dislocation $\alpha\delta$ has the Burgers vector $\frac{1}{6}[101]$. The glissile Schockley-partial δB slips along the twin boundary, resulting in untwinning on one layer of the twin lamella. If many initial dislocations slip on successive planes and they meet the twin lamella, each dislocation causes untwinning on a layer in the lamella that leads to a complete disappearance of a segment of the twinned region (see Figure 3.11b). Then, further dislocations can pass freely.

It should be noted that dynamic recovery mechanisms may also contribute to the decrease of the dislocation density in Ag between 8 and 16 passes in a manner similar to that shown for Cu processed by more than 8 passes in ECAP [22, 23]. The high stresses at twin tips may assist the cross-slip of dissociated dislocations in Ag, which plays an important role in recovery, whereas for Cu twinning remains marginal even when the ECAP processing is continued to extremely high strains [22, 23]. However, for the case of Ag where SFE is very low, the influence of twinning on the dislocation density must also be considered, as discussed earlier. It is noted that the evolution of the defect structure is affected by the impurity level. In UFG or nanomaterials, the segregation of alloying elements at the grain boundaries may reduce the activity of these boundaries as nucleation sites of twin lamellas, since the impurities hinder the emission of twining partials.

3.3 Effect of low stacking fault energy on defect structure in ultrafine-grained alloys

In some binary systems (e.g. in Cu-Zn or Cu-Al), the alloying is accompanied by the decrease of SFE [24–26]. Therefore, the influence of low SFE on the defect structure in UFG alloys can be studied by increasing the solute atom concentration in these alloys. This effect is illustrated on nanostructured Cu, Cu-10% Zn and Cu-30% Zn alloys having SFEs of about 62, 35 and 10 mJ m^{-2}, respectively [27, 28]. The samples were processed by HPT for five revolutions at RT. After five revolutions of HPT, the dislocation density values at the half-radius of the as-processed disks are 14 ± 3, 34 ± 5 and $81 \pm 7 \times 10^{14}$ m^{-2} for 99.9% purity Cu, Cu-10% Zn and Cu-30% Zn, respectively [27]. The corresponding crystallite size values are 60 ± 5, 60 ± 5 and 34 ± 3 nm for 99.9% purity Cu, Cu-10% Zn and Cu-30% Zn, respectively [27]. The significant increase of dislocation density and reduction in crystallite size with increasing Zn concentration are caused partially by the effect of alloying and partially by the decrease of SFE. The smaller the SFE, the larger the degree of dislocation dissociation and therefore the dislocation annihilation is more retarded, as discussed above in the case of pure fcc metals. The dislocation arrangement parameter, M, obtained by x-ray line profile analysis, increases with increasing Zn content in Cu [26]. This quantity characterises the degree of screening of strain fields of dislocations: the stronger screening corresponds to smaller value of M [29]. The larger M for higher Zn concentration suggests that the arrangement of dislocations into low energy configurations is hindered by the difficult cross-slip and climb due to large solute atom content and low SFE.

It is noted that the crystallite size and the dislocation density remain unchanged within the experimental error along the radius of the disks for samples Cu and Cu-30% Zn while the crystallite size decreases and the dislocation density increases with increasing the distance from the centre for specimen Cu-10% Zn [26]. The average grain sizes as obtained from TEM images are 345 nm, 240 nm and

160 nm in the centres for Cu, Cu-10% Zn and Cu-30% Zn, respectively [27]. The corresponding values in the periphery of the disks are 75, 50 and 10 nm [27]. The latter value is extremely small among the SPD-processed metals. In both the centre and the periphery of the disks, the grain size decreases with increasing Zn content. It should be noted that in Cu-Zn alloys the effects of alloying and small SFE are superimposed, therefore it is difficult to separate their contributions in the increase and decrease of the dislocation density and the grain size, respectively.

While in UFG Cu twins are scarcely observed, the smaller SFE in Cu-10% Zn and Cu-30% Zn alloys results in significant twinning during HPT [27]. For Cu-10% Zn the twin boundary frequency values are 0.16%, 0.82% and 1.36% at the centre, half-radius and periphery of the disk, respectively [27]. The corresponding values for Cu-30% Zn are 1.82%, 2.88% and 3.73%, respectively [27]. In UFG or nanostructured Cu-Al alloys processed by SPD, the increment of twin density with increasing Al content was also observed, which can be also explained by the decrease of SFE [26, 30]. For Cu-5 at.% Al, Cu-8 at.% Al and Cu-16 at.% Al alloys, the values of SFE are 28 mJ m^{-2}, 17 mJ m^{-2} and 6 mJ m^{-2}, respectively [26]. The twin boundary frequency achieved by SPD for some nanostructured low SFE metals and alloys are listed in Table 3.2. In most cases, the twin boundary frequency is determined by x-ray line profile analysis. When x-ray data are not available, the percentage twin boundary frequency (β) is estimated from the mean spacing (d_{twin}) between twin boundaries in TEM images as:

$$\beta = \frac{100 \cdot d_{111}}{d_{twin}} \tag{3.15}$$

where d_{111} is the distance between the neighbouring {111} planes. The twin boundary frequency data in Table 3.2 are

Table 3.2 The maximum values of the twin boundary frequency achieved by SPD for some nanostructured low SFE metals and alloys. RT: RT. XRD or TEM indicate that the twin boundary frequency values are determined by line profile analysis of x-ray diffraction peaks or from the mean twin-spacing obtained by TEM. The values of SFE and grain size are also listed

Material	Processing method	SFE [mJ/m^2]	Grain size [nm]	Twin boundary frequency [%]	Reference
Au	4 ECAP at RT	38 [2, 3]	460	0.3 (XRD)	[33]
Ag	16 ECAP at RT	19 [2, 3]	200	1.5 (XRD)	[12]
Cu-10% Zn	5 HPT at RT, disk centre	35 [22]	240	0.16 (XRD)	[27]
Cu-10% Zn	5 HPT at RT, disk periphery	35 [22]	50	1.36 (XRD)	[27]
Cu-30% Zn	5 HPT at RT, disk centre	10 [22]	160	1.82 (XRD)	[27]
Cu-30% Zn	5 HPT at RT, disk periphery	10 [22]	10	3.73 (XRD)	[27]
Cu-5% Al	4 ECAP at RT	28 [24]	107	~1 (TEM)	[26]
Cu-8% Al	4 ECAP at RT	17 [24]	82	~2 (TEM)	[26]
Cu-16% Al	2 ECAP at RT	6 [24]	50	~2 (TEM)	[26]

Figure 3.12 The twin boundary frequency as a function of SFE. The grain size is also given at some data points. The data are taken from Table 3.2

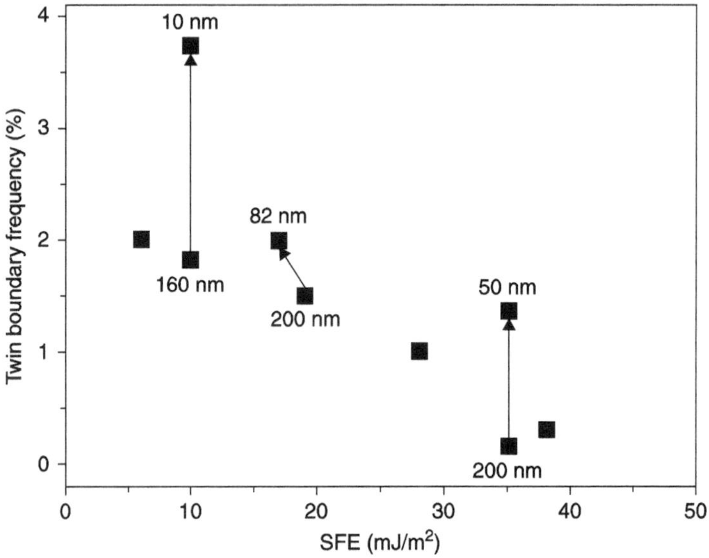

plotted as a function of SFE in Figure 3.12. It is revealed that the large twin boundary frequency is basically caused by the low SFE. At the same time, for same or similar SFE values, the smaller grain size is accompanied by larger twin boundary frequency. It is noted that in medium or high SFE metals (e.g. in Cu or Al) twinning is also observed if the grain size is very small (about 20 nm) [31, 32]. The effect of grain size on twinning activity is presented in Chapter 5.

3.4 Grain-refinement mechanisms in low stacking fault energy alloys

In the case of materials having medium or high SFE, dislocation activities play a major role in the grain-refinement during SPD procedures. During SPD processes, dislocations are formed, interact with each other and rearrange into cell- and subgrain-boundaries. With increasing strain in SPD, the misorientation between neighbouring subgrains increases and they become high-angle grain boundaries, thereby resulting in a refinement of the initial coarse grains. Simultaneously, new subgrain boundaries are formed by dislocation accumulation in the refined grains, which yields further grain-refinement. For materials with low SFEs, it is expected that stacking faults and twin boundaries formed during deformation play a significant role in grain-refinement. The strong effect of SFE on the grain-refinement mechanisms and the minimum achievable grain size is demonstrated on Cu-30% Zn alloy having extremely low SFE of about 10 mJ/m^2, which was subjected to five revolutions by HPT at RT [34]. As the imposed strain is linearly proportional to the distance from the centre in HPT-processed disks, the different stages of microstructural evolution were studied by selecting various positions on the disk processed by five revolutions.

The steps of grain-refinement as observed in TEM images are depicted in the schematic drawing of Figure 3.13 [34]. In the first step, the initial equiaxed ultrafine grains with grain sizes of hundreds of nanometres are divided into elongated regions having widths of about 10 nm through the formation of a high density of stacking faults and twins. Most of the stacking faults and twins are formed through partial dislocation emission from grain boundaries and some of them terminate in grain interiors [34]. A few twins may form through a dissociation of lattice dislocation into two partials as both ends terminate within a grain.

In step 2 (see Figure 3.13), a high density of dislocations accumulates at stacking faults and twin boundaries, therefore the original atomically flat coherent stacking faults and twin boundaries evolve into curved incoherent high angle grain

Figure 3.13 **Schematic illustration of the grain-refinement mechanism for the Cu-30 wt.% Zn alloy processed by HPT**

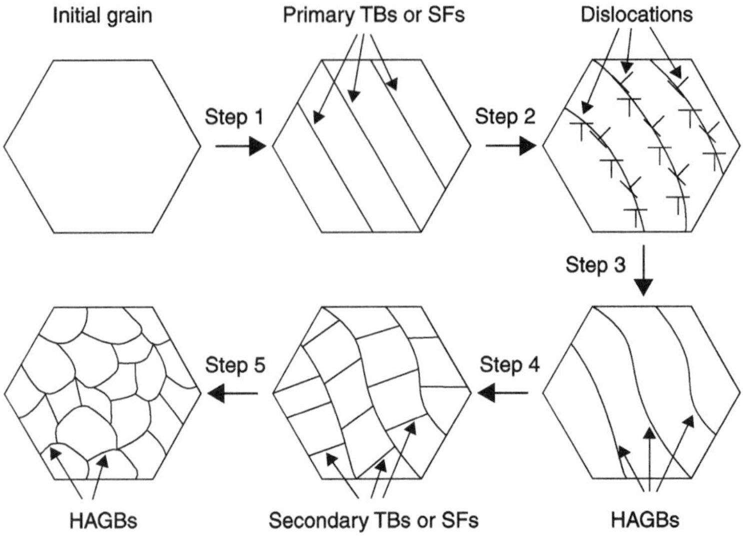

Source: After Wang et al. (2010) [34] with permission from Elsevier.

boundaries (step 3). The HRTEM image in Figure 3.14 shows dislocations blocked at a twin boundary resulting in an 8° deviation from the ideal misorientation between twins [34]. In step 4, secondary stacking faults and twins were emitted from these boundaries as illustrated in Figure 3.13 [34]. Finally, in step 5 the emissions of secondary stacking faults and twins further refine the grains and transform the elongated grains into equiaxed grains having a grain size of 10 nm. The experimental observations presented above suggest that both dislocations and twin boundaries participate in the grain-refinement during SPD of low SFE materials. First, twin boundaries divide the initial grains into smaller parts, and then these twin boundaries will be transformed into incoherent grain boundaries by accumulation of gliding dislocations. The final grain size of 10 nm is extremely low among the

Figure 3.14 A high-resolution TEM image of a bent TB in HPT-processed Cu-30% Zn alloy showing a high density of dislocations, which are indicated using white and black 'T', accumulated at the TB. Two white solid lines were drawn parallel to {111} at each side of the TB, respectively, and one dashed white line was also drawn parallel to $(111)_M$ to indicate the misorientation between the {111} planes across the TB

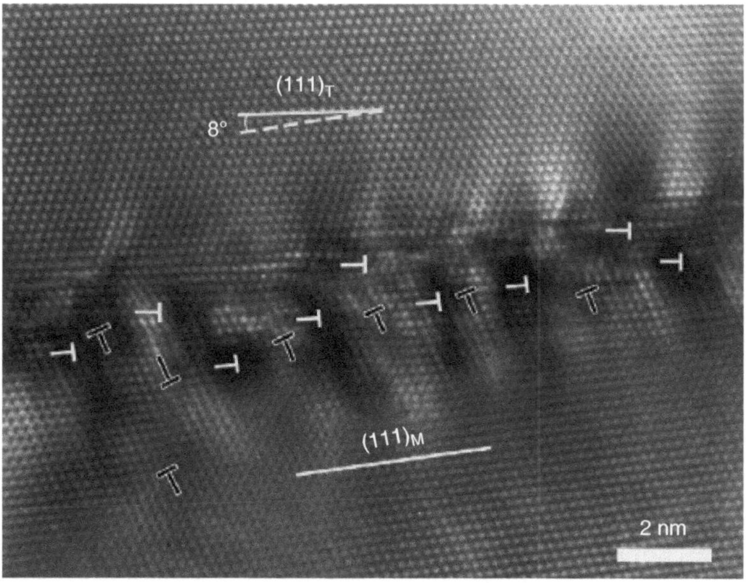

Source: Reprinted from Wang et al. (2010) [34] with permission from Elsevier.

HPT-processed metals, which can also be attributed to the effect of high alloying element concentration in addition to the low SFE as the solute atoms hinder the recovery and recrystallisation of the nanostructure. It is worth noting that an amorphous structure was also formed at the incoherent boundaries of the final nanocrystalline grains [34].

Similar effect of low SFE on grain-refinement mechanisms can be observed in Cu–Al alloys processed by ECAP at RT

[26]. With increasing the Al content in Cu–Al alloys, the SFE decreases [26]. For Cu-5 at.% Al alloy, although the grain-refinement is dominated by the dislocation activity, twinning also has a contribution to the decrease of grain size. The increase of Al concentration yields a gradual transformation of the ECAP-induced grain-refinement mechanism from dislocation subdivision to twin fragmentation [26]. Similarly to HPT-processed Cu-30% Zn alloy, primary and secondary twin boundaries are formed after the second ECAP pass and they facilitate the grain-refinement. The grain size values after four passes for Cu-5 at.% Al and Cu-8 at.% Al alloys are 107 and 82 nm, respectively, while the grain size for Cu-16 at.% Al after two passes is 50 nm [26]. These values are very low among the ECAP-processed materials, which can be attributed to the combined effect of alloying and low SFE as discussed above. It is worth noting that the HPT-processed Cu-30% Zn alloy has much smaller grain size (10 nm) than Cu-16 at.% Al alloy after two passes of ECAP despite their very similar SFE (about 10 mJ m^{-2}). This difference may be caused by the larger solute atom concentration in Cu-30% Zn alloy. The effectiveness of twinning mechanism in grain-refinement is also manifested in the much higher fraction of high-angle grain boundaries in Cu-8 at.% Al alloy (62%) compared with Cu-5 at.% Al alloy (49%) [26].

3.5 References

1. I. Kovács and L. Zsoldos, *Dislocations and Plastic Deformation*, Pergamon Press, London, 1973.
2. J.P. Hirth and J. Lothe, *Theory of Dislocations*, Wiley, New York, 1982.
3. L.E. Murr, *Interfacial Phenomena in Metals and Alloys*, Addison Wesley, Reading, MA, 1975.

4. X.Z. Liao, S.G. Srinivasan, Y.H. Zhao, M.I. Baskes, Y.T. Zhu, F. Zhou, E.J. Lavernia and H.F. Xu, 'Formation mechanism of wide stacking faults in nanocrystalline Al', *Applied Physics Letters*, 84 (2004) 3564–3566.
5. X.Z. Liao, F. Zhou, E.J. Lavernia, S.G. Srinivasan, M.I. Baskes, D.W. He and Y.T. Zhu, 'Deformation mechanism in nanocrystalline Al: Partial dislocation slip', *Applied Physics Letters*, 83 (2003) 632–634.
6. J. Friedel, *Dislocations and Mechanical Properties of Crystals*, John Wiley and Sons, New York, 1957.
7. P.B. Escaig, 'Sur le glissement dévié des dislocations dans la structure cubique à faces centrées', *Journal de Physique*, 29 (1968) 225–239.
8. A.N. Stroh, 'The formation of cracks as a result of plastic flow', *Proceedings of the Royal Society*, A 2243 (1954) 404–414.
9. G. Saada, 'Internal stresses, dislocation mobility and ductility', *Chinese Journal of Physics*, III 1 (1991) 945–956.
10. J.K. Baria, 'Temperature dependent lattice mechanical properties of some fcc transition metals', *Chinese Journal of Physics*, 42 (2004) 287–306.
11. A.S. Argon and W.C. Moffatt, 'Climb of extended edge dislocations', *Acta Metallurgica*, 29 (1981) 293–299.
12. J. Gubicza, N.Q. Chinh, J.L. Lábár, Z. Hegedűs and T.G. Langdon, 'Principles of self-annealing in silver processed by equal-channel angular pressing: The significance of a very low stacking fault energy', *Materials Science and Engineering*, A 527 (2010) 752–760.
13. N.Q. Lam, S.J. Rothman, H. Mehrer and L.J. Nowicki, 'Self-diffusion in silver at low temperatures', *Physica Status Solidi*, (b) 57 (1973) 225–236.

14. R.E. Hoffmann and D. Turnbull, 'Lattice and Grain Boundary Self-Diffusion in Silver', *Journal of Applied Physics*, 22 (1951) 634–639.
15. J. Gubicza, N.Q. Chinh, J.L. Lábár, Z. Hegedűs, C. Xu and T.G. Langdon, 'Microstructure and yield strength of severely deformed silver', *Scripta Materialia*, 58 (2008) 775–778.
16. P. Müllner and C. Solenthaler, 'On the effect of deformation twinning on defect densities', *Materials Science and Engineering*, A 230 (1997) 107–115.
17. J.A. Venables, 'Deformation twinning in face-centred cubic metals', *Philosophical Magazine*, 6 (1961) 379–396.
18. J.B. Cohen and J. Weertman, 'A dislocation model for twinning in fcc metals', *Acta Metallurgica*, 11 (1963) 996–998.
19. S. Mahajan and G.Y. Chin, 'Formation of deformation twins in fcc crystals', *Acta Metallurgica*, 21 (1973) 1353–1363.
20. Y.T. Zhu, X.Z. Liao, S.G. Srinivasan and E.J. Lavernia, 'Nucleation of deformation twins in nanocrystalline face-centered-cubic metals processed by severe plastic deformation', *Journal of Applied Physics*, 98 (2005) 034319.
21. Z.W. Wang, Y.B. Wang, X.Z. Liao, Y.H. Zhao, E.J. Lavernia, Y.T. Zhu, Z. Horita and T.G. Langdon, 'Influence of stacking fault energy on deformation mechanism and dislocation storage capacity in ultrafine-grained materials', *Scripta Materialia*, 60 (2009) 52–55.
22. J. Gubicza, N.Q. Chinh, J.L. Lábár, S. Dobatkin, Z. Hegedűs and T.G. Langdon, 'Correlation between microstructure and mechanical properties of severely deformed metals', *Journal of Alloys and Compounds*, 483 (2009) 271–274.

23. F. Dalla Torre, R. Lapovok, J. Sandlin, P.F. Thomson, C.H.J. Davies and E.V. Pereloma, 'Microstructures and properties of copper processed by equal channel angular extrusion for 1–16 passes', *Acta Materialia*, 52 (2004) 4819–4832.
24. J.M. MacLaren, A. Gonis and G. Schadler, 'First-principles calculation of stacking-fault energies in substitutionally disordered alloys', *Physical Review*, B 45 (1992) 14392–14395.
25. Y.H. Zhao, Z. Horita, T.G. Langdon and Y.T. Zhu, 'Evolution of defect structures during cold rolling of ultrafine-grained Cu and Cu-Zn alloys: Influence of stacking fault energy', *Materials Science and Engineering*, A 474 (2008) 342–347.
26. S. Qu, X.H. An, H.J. Yang, C.X. Huang, G. Yang, Q.S. Zang, Z.G. Wang, S.D. Wu and Z.F. Zhang, 'Microstructural evolution and mechanical properties of Cu–Al alloys subjected to equal channel angular pressing', *Acta Materialia*, 57 (2009) 1586–1601.
27. L. Balogh, T. Ungar, Y. Zhao, Y.T. Zhu, Z. Horita, C. Xu and T.G. Langdon, 'Influence of stacking-fault energy on microstructural characteristics of ultrafine-grain copper and copper-zinc alloys', *Acta Materialia*, 56 (2008) 809–820.
28. Z.J. Zhang, Q.Q. Duan, X.H. An, S.D. Wu, G. Yang and Z.F. Zhang, 'Microstructure and mechanical properties of Cu and Cu–Zn alloys produced by equal channel angular pressing', *Materials Science and Engineering*, A 528 (2011) 4259–4267.
29. T. Ungár, J. Gubicza, G. Ribárik and A. Borbély, 'Crystallite size distribution and dislocation structure determined by diffraction profile analysis: principles and practical application to cubic and hexagonal crystals', *Journal of Applied Crystallography*, 34 (2001) 298–310.

30. V. Subramanya Sarma, J. Wang, W.W. Jian, A. Kauffmann, H. Conrad, J. Freudenberger and Y.T. Zhu, 'Role of stacking fault energy in strengthening due to cryo-deformation of fcc metals', *Materials Science and Engineering*, A 527 (2010) 7624–7630.
31. X.Z. Liao, F. Zhou, E.J. Lavernia, D.W. He and Y.T. Zhu, 'Deformation twins in nanocrystalline Al', *Applied Physics Letters*, 83 (2003) 5062–5064.
32. X.Z. Liao, Y.H. Zhao, S.G. Srinivasan, Y.T. Zhu, R.Z. Valiev and D.V. Gunderov, 'Deformation twinning in nanocrystalline copper at room temperature and low strain rate', *Applied Physics Letters*, 84 (2004) 592–594.
33. J. Gubicza, N.Q. Chinh, P. Szommer, A. Vinogradov and T.G. Langdon, 'Microstructural characteristics of pure gold processed by equal-channel angular pressing', *Scripta Materialia*, 56 (2007) 947–950.
34. Y.B. Wang, X.Z. Liao, Y.H. Zhao, E.J. Lavernia, S.P. Ringer, Z. Horita, T.G. Langdon and Y.T. Zhu, 'The role of stacking faults and twin boundaries in grain refinement of a Cu–Zn alloy processed by high-pressure torsion', *Materials Science and Engineering*, A 527 (2010) 4959–4966.

4

Defects in nanomaterials processed by powder metallurgy

Abstract: In this chapter, the type and densities of lattice defects in nanopowders and bulk materials consolidated from these powders are overviewed. The evolutions of arrangement, edge/screw character and density of dislocations, and grain size during milling of metallic powders are presented. The minimum grain size achievable by milling at RT is correlated to the melting point of metals. The effects of the initial powder particle size and the consolidation conditions on the microstructure of sintered metals, diamond and ceramics are discussed. The powder consolidation at high pressures and temperatures usually yields grain-growth, annihilation of twin boundaries and increase in the dislocation density. When blends of nano- and coarse-grained powders are sintered, the defect structure in the nanocrystalline fraction of the consolidated material is strongly influenced by the amount of the coarse-grained powder component and vice versa.

Key words: powder metallurgy, milling, dislocation density, twinning, minimum grain size, grain growth during consolidation.

4.1 Development of defect structure during milling

Ball milling is a top-down approach for the production of nanopowders. Depending on the chemical composition of the initial powder and on the conditions in milling, different processes can occur, such as grain-refinement, mechanochemical activation synthesis, mechanical alloying, reactive mechanical alloying, mechanical synthesis, mechanical disordering or mechanical amorphisation [1]. In addition to grain-refinement, short milling times can break the thin, chemically passive surface coatings (e.g., surface oxides), resulting in an increased chemical activity of the milled media toward both gasses and chemical reactions in solutions and electrolytes. This process is often termed *mechanochemical activation synthesis*. In *mechanical alloying*, blends of powders of different elements are milled together to trigger alloying of elemental powders in a solid-state process. Since long-distance migration of chemical species is necessary for alloying, this process requires long times of milling, usually more than twice the time needed for reaching the minimum grain size achievable at the given conditions. When mechanical alloying initiates a solid-state reaction and yields a new stoichiometric or quasi-stoichiometric chemical compound, such as carbide or nitride, the process can be termed *reactive mechanical alloying*. When the milling of two chemical compounds yields to a distinct compound, the process is termed *mechanical synthesis*. Long milling of intermetallics leads to disorder in the structure; therefore this process is termed *mechanical disordering*. Long milling may also result in *mechanical amorphisation* of crystalline powders [1].

The evolution of microstructure in mechanical alloying is illustrated with the case of milling blends of Al and Mg powders. A series of Al-Mg samples were prepared from high purity aluminium (99.9%) powder and high purity magnesium chips (less than 2 mm thick and 5 mm long) [2]. The mechanical alloying was carried out using a shaker mill at room temperature (RT). Since both the balls and the vial of the mill are made from hardened steel, the iron contamination in the soft Al alloy is very small, below 0.02 at.%. Powder blends of Al and 6 wt.% Mg were milled for 0.5, 1, 3 and 6 h. Another series of Al – x wt.% Mg (x = 0, 3, 6) were milled for 3 h at the same conditions to study the effect of the nominal Mg content on the evolution of microstructure. The Mg solute concentrations in Al were determined from the lattice constants measured after milling. As can be seen from Figure 4.1, when Al and 6 wt.% Mg powders are milled together, the Mg concentration in Al

Figure 4.1 The mean crystallite size, the dislocation density and the Mg concentration as a function of the milling period for the powder blend with the nominal composition of Al–6 wt.% Mg

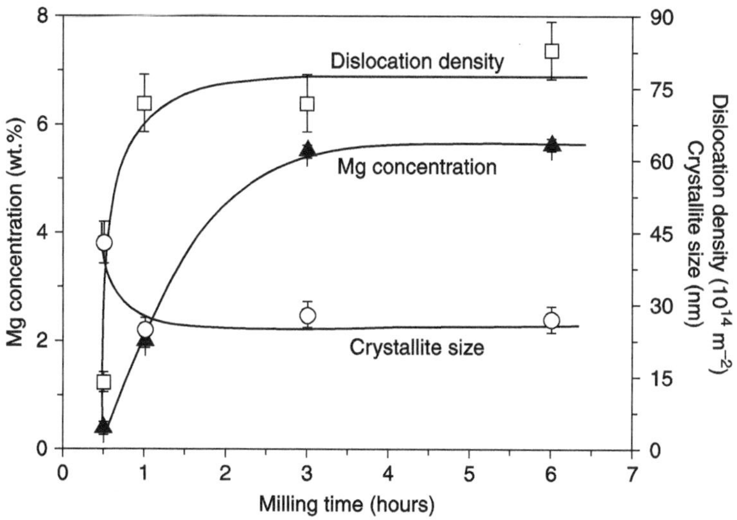

increases with milling time up to 3 h [2]. After milling for 3 h, almost all of the Mg content in the powder blend goes into solid solution (the Mg concentration is 5.52 wt.%). The solubility limit of Mg in Al at RT is lower than 1.9 wt.% [3, 4]. Figure 4.1 shows that this value is well exceeded by mechanical alloying as a result of the non-equilibrium state of the ball milled alloys [5, 6]. The Mg atoms are larger than Al, therefore they are located preferably at lattice defects, such as core of edge dislocations or grain boundaries. During milling, a large amount of lattice defects develops, which enables the accommodation of Mg atoms with larger concentration than the equilibrium value. The collision between the milling balls and the powder yields cold welding of Al and Mg particles, which enables powder particles to be in contact, thereby facilitating diffusion of Mg atoms into Al lattice. Welding is followed by fracturing and rewelding at another location on the particles' surfaces in the course of milling. The fracturing of the welded particles impedes their clustering, which promotes the transfer of the high ball collision energy to all particles and produces new, clean surfaces without oxide layers accelerating the diffusion. In milling procedures, control agents are used to avoid excessive cold welding and oxidation of powder particles.

The mean crystallite size and the dislocation density obtained by x-ray line profile analysis as a function of the milling time are plotted in Figure 4.1. The dislocation density increases while the crystallite size decreases rapidly with milling time up to 1 h [2]. Between 1 and 6 h milling time, the dislocation density and the crystallite size do not change significantly. It should be noted that the crystallite size or the grain size in metallic powders processed by milling is usually smaller by two to three orders of magnitude than the powder particle size, as depicted in Figure 4.2. This suggests that the final nanograins form rather inside the particles by similar

Figure 4.2 **Schematic illustration of the particle/grain structure in milled metallic powders**

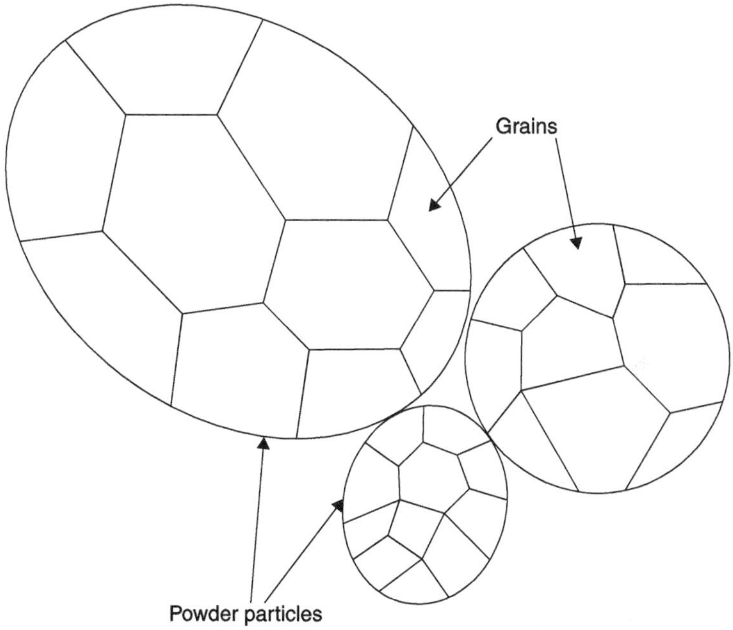

mechanisms as in severe plastic deformation (SPD) of bulk metals (see Chapter 2) than by the fracture of particles into nanosized fragments. However, the latter phenomenon may occur during milling of brittle materials, such as ceramics, and then the crystallite size is close to the particle size. The saturation of Mg solute concentration in Al is reached at longer milling time than the maximum value of dislocation density and the minimum value of crystallite size as the migration of Mg atoms from the surface to the interior of particles by diffusion at RT requires longer time. It is also worth noting that the evolution of the dislocation structure does not finish after the saturation of dislocation density as it is indicated by the change of the dislocation arrangement parameter *M* determined by x-ray line profile analysis (see

Figure 4.3). The dimensionless parameter *M* is defined as the product of the outer cut-off radius of dislocations and the square-root of the dislocation density. The value of parameter *M* decreases with increasing milling time up to 3 h, which indicates an enhancement in screening of the strain field of dislocations due to their arrangement into low energy configurations, such as dipolar walls or low-angle grain boundaries [2]. This process continues even after the saturation of the dislocation density, i.e. between 1 and 3 h of milling. After 3 h milling time, parameter *M* does not change. The crystallite size and the dislocation density follow similar trends with increasing milling time when the milling is performed on pure Al, i.e. alloying does not occur [6].

The edge/screw character of dislocations can be obtained from parameter *q* determined by x-ray line profile analysis.

Figure 4.3 **The parameters *q* and *M* describing the edge/screw character and the screening of dislocations, respectively, as a function of milling time for a powder blend with the nominal composition of Al–6 wt.% Mg**

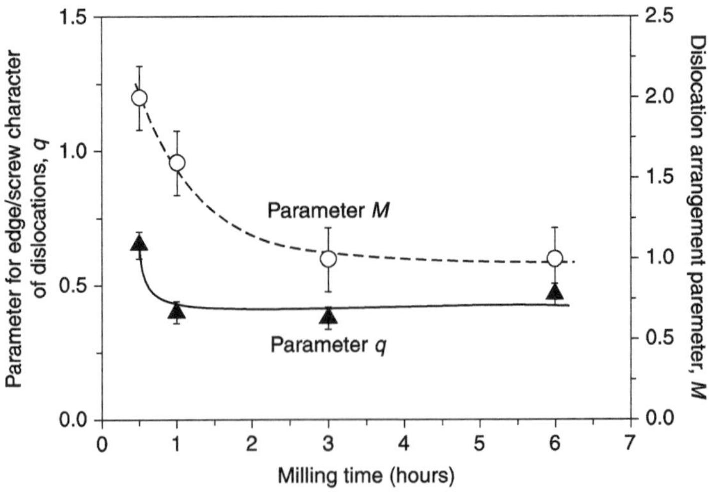

The values of parameter q in Al for pure screw and edge dislocations in the <110>{111} slip system are 1.33 or 0.36, respectively [7]. The evolution of q in Al milled together with 6 wt.% Mg as a function of milling time is shown in Figure 4.3. The values of q are lower than the arithmetic average of pure edge and screw cases, 0.85, which indicates that the character of dislocations is rather edge type. The edge character of dislocations becomes stronger with increasing milling time, which can be explained by the high stacking fault energy (SFE) of Al. The large value of SFE yields much easier annihilation of screw dislocations by cross-slip than edge dislocations by climb during SPD, as discussed in Chapter 3. Additionally, with increasing milling time, the solute Mg concentration in Al increases, and the pinning effect of Mg atoms is more effective on edge than screw dislocations. Mg atoms have a larger atomic radius than Al atoms, therefore they are preferentially diffusing to the dilatation side of edge dislocations [8], leaving screw dislocations unpinned. Due to this effect screw dislocations are annihilated more easily during the milling procedure.

Figure 4.4 illustrates the evolution of the Mg solute atom concentration in Al with increasing the nominal Mg content in Al-Mg powder blend for fixed milling time of 3 h. The Mg concentration in Al increases with the amount of Mg in the blends and, after 3 h milling, more than 75% of the Mg content is dissolved in Al [2]. The dislocation density increases while the crystallite size decreases with increasing the nominal Mg content, as shown in Figure 4.4, which can be attributed to the pinning effect of Mg solute atoms on dislocations and subgrain/grain boundaries. Figure 4.5 illustrates that the stronger pinning effect of Mg atoms on edge dislocations compared with screw dislocations yields a shift of dislocation character toward edge type with increasing the nominal Mg content in the samples milled for

Figure 4.4 The mean crystallite size, the dislocation density and the Mg concentration in Al as a function of the nominal Mg content in Al-Mg powder blends milled for 3 h

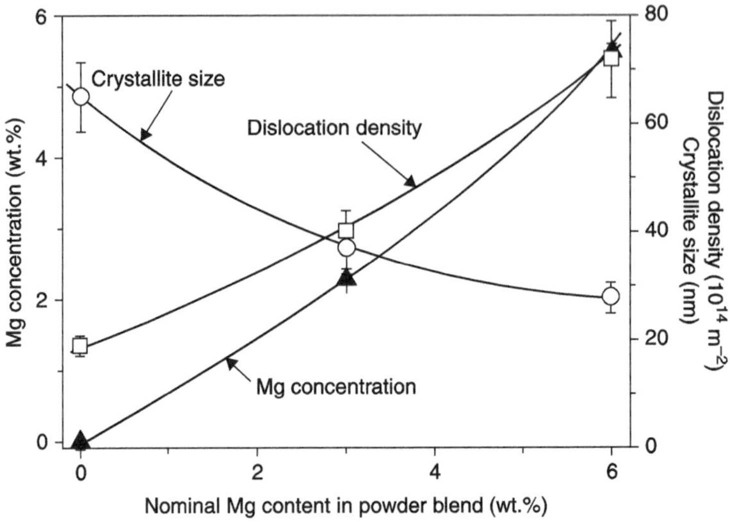

Figure 4.5 The parameters q and M describing the edge/screw character and the screening of dislocations, respectively, as a function of the nominal Mg content in Al-Mg powder blends milled for 3 h

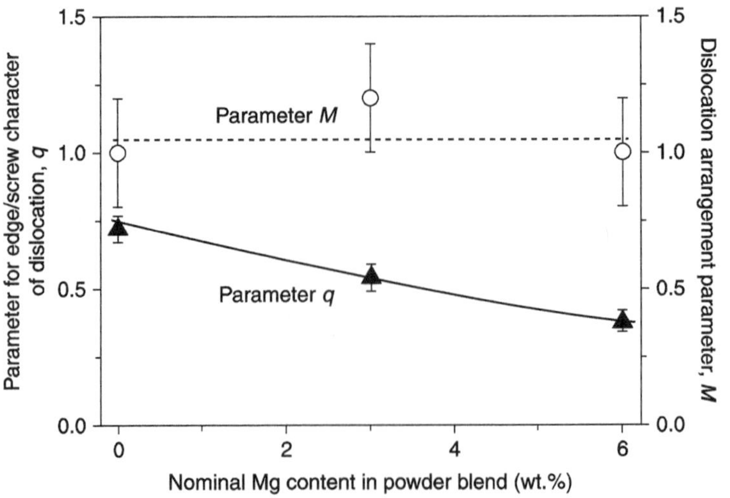

3 h. At the same time, the value of parameter M remained unchanged within the experimental error when the nominal Mg content increases. This indicates that if the milling time is long enough, the arrangement of dislocations is only slightly affected by the Mg solute concentration. Although the strong interaction between solute Mg atoms and dislocations decreases dislocation mobility [9, 10], this effect may be overwhelmed after sufficiently long time of milling due to the action of high stresses induced by SPD.

In a milled powder, the larger particles usually contain larger crystallites, as illustrated in Figure 4.2. For instance, in the case of a 99.8% purity Al powder milled in liquid nitrogen at a temperature of ~70 K for 25 h (cryomilling), the particle size spans between a few tens of micrometres and 150 μm while the crystallite sizes are only between 30 nm and 120 nm [11]. The grain size determined by transmission electron microscopy (TEM) and the crystallite size obtained by x-ray line profile analysis are in good agreement. Despite the three orders of magnitude difference between the crystallite and particle sizes, a nearly linear relationship between the two quantities has been observed. The smaller particles contain not only smaller crystallites, but also higher dislocation density. The higher dislocation density and smaller crystallite size in finer particles can be attributed to the larger concentration of impurities. The contamination of a milled powder is mainly caused by the interaction between the particles and the milling media. For cryomilled Al powders, the most probable impurity is nitrogen. It may segregate at grain boundaries as solute clusters, and also may cause precipitation of AlN nanoparticles, if the powder is heated up to high temperatures [11]. Additionally, the alumina and aluminium-hydroxide phases on the surface of Al particles may break during milling and be incorporated by the severely deformed Al particles. Both the solute impurities and the secondary phase dispersoids hinder the annihilation processes of lattice defects. Since a finer particle has a larger specific surface area, both the surface alumina fraction and the nitrogen concentration migrating through the surface are larger, which yield higher dislocation density and smaller crystallite size inside the particle [11].

Experiments conducted on pure metallic powders revealed that the milling conditions have a significant effect on the defect structure evolved in the particles. It is suggested for a constant milling time in a planetary mill that the crystallite size and the parameters of the lattice defect structure can be correlated to the total collision power and the ratio of the shocked and friction kinetic energies [12, 13]. The total collision (or shock) power is calculated as the product of the frequency of collisions and the total kinetic energy (E_t) of a shock. As the angle between the collision velocity vector and the vial surface depends on the ball milling condition, the total kinetic energy can be decomposed into two components: the friction energy and the shock energy. To obtain a quantitative value of the friction energy and the shock energy, the collision velocity is decomposed into two components: a normal velocity and a tangential velocity. The normal velocity component is perpendicular to the vial surface and induces the shock energy (E_s), while the tangential velocity component is parallel to the vial surface and induces the friction energy (E_f) [14]. In a planetary mill, the vials containing the powder and the balls are fixed onto a disc. For given mill geometry and balls parameters (size, mass, number), the shock energy and the ratio E_s/E_t can be varied independently by changing the rotation speeds of the vials and the disc [13]. It was found for Fe powder milled at RT for 24 h that the crystallite size decreases and the dislocation density increases with increasing shock power. In addition to the shock power, the ratio E_s/E_t also influences the evolution of the dislocation structure as the nature of dislocations shifts from mainly edge to mainly screw type in Fe with increasing this ratio [12]. The screw character of dislocations in Fe powder also becomes stronger with increasing milling time at RT [15]. This can be explained by the reduced mobility of screw dislocations in bcc structures due to dissociation of their cores into non-planar configurations [16]. As a consequence, during milling the edge dislocation segments can annihilate more easily than the screw ones, therefore the remaining dislocations have more screw character.

In the case of Cu powder milled for a constant time, the evolutions of the crystallite size and the dislocation density are controlled by the shock energy, whereas the edge/screw character of dislocations depends on E_s/E_t [13]. The crystallite size decreases and the dislocation density increases with increasing shock energy, while the edge character of the dislocation structure becomes stronger when E_s/E_t increases. The latter phenomenon can be explained by the easier cross-slip of screw dislocations than climb of edge segments due to the low degree of planar dissociation of dislocation cores (see Chapter 3). In ball-milled Cu powders, considerable amounts of twin boundaries are detected. The maximum twin boundary frequency determined by x-ray line profile analysis in Cu powders milled at RT is 4.4% [13]. The twin boundary frequency in face-centred cubic (fcc) structures is defined as the relative fraction of twin boundaries among the {111} lattice planes. The twin boundary frequency shows a non-monotonous evolution as a function of shock energy: in the first stage it increases, then in the second stage it decreases and finally in the third stage it increases again. The increment in twinning activity in the third stage is accompanied by the reduction of the dislocation density. The origin of this non-monotonous behaviour of twin boundary frequency is not yet clear, but its increase in the first stage can be attributed to deformation twinning caused by the large number of glide obstacles formed due to high dislocation density. In the third stage, the higher twin boundary frequency at the expense of dislocations can be related to the modification of plastic deformation mechanism from dislocation motion to twinning in very small grain sizes as predicted by modelling of plastic deformation mechanism in nanomaterials [17]. The change of deformation mechanisms with decreasing grain size is discussed in Chapter 5.

The maximum dislocation density and the minimum crystallite size achieved in different powder materials (pure metals, solid solutions, dispersion-strengthened alloys and compounds) by milling are determined by x-ray line profile analysis and listed in Table 4.1. In some cases, the grain size is determined by TEM. Despite the similar grain refinement mechanisms as in SPD-processing of bulk materials, milling results in much higher maximum dislocation density and smaller crystallite and grain sizes. Additionally, for the samples where both the grain and crystallite sizes are available, a good agreement between the two quantities can be established [18–21], while in bulk SPD-processed materials the grain size is usually two to six times larger than the crystallite size (see Chapter 2). These deviations can be explained by the higher production rate of dislocations due to the very severe deformation between the balls and the particles, and also by the contamination introduced into the powder during milling. The minimum grain size achievable by milling is attributed to a balance between dislocations introduced by the plastic deformation of milling and their recovery by thermal processes [22]. The retarding effect of impurities on the annihilation of lattice defects yields higher maximum dislocation density and smaller crystallite size than the values obtained

Table 4.1 The maximum dislocation density and the minimum crystallite size determined by x-ray line profile analysis, and the minimum grain size obtained by TEM for powder materials processed by long time milling. The temperature of milling was RT except for the cases indicated by asterisks, where the milling was performed at liquid nitrogen temperature. For pure metals, the melting point values are also given in order to reveal their correlation to the minimum crystallite size (see Figure 4.6)

Material	Melting point [K]	Grain size [nm]	Crystallite size [nm]	Dislocation density [10^{14} m^{-2}]	Reference
Pure metals					
Al	933	–	22	–	[23, 24]
Al*	933	28	26	–	[19]
Al	933	–	39	40	[25]
Ag	1,235	–	25	–	[24]
Cu	1,356	–	21	–	[23, 24]
Cu	1,356		35	740	[13]
Rh	2,236	–	7	–	[23, 24]
Ir	2,720	–	6	–	[23, 24]
Ni	1,726	–	12	–	[23, 24]
Ni	1,726	7	8	240	[26]
Pd	1,825	–	7	–	[23, 24]

(Continued)

Table 4.1 The maximum dislocation density and the minimum crystallite size determined by x-ray line profile analysis, and the minimum grain size obtained by TEM for powder materials processed by long time milling (*Continued*)

Material	Melting point [K]	Grain size [nm]	Crystallite size [nm]	Dislocation density [10^{14} m^{-2}]	Reference
W	3,695	–	7	–	[24]
Fe	1,808	–	8	–	[24]
Fe	1,808	–	15	2,980	[12]
Fe	1,808	–	19	165	[27]
Cr	2,130	–	9	–	[24]
Hf	2,742	–	12	–	[24]
C (diamond)	3,825	–	3	–	[24]
Si	1,683	–	8	–	[24]
Nb	2,742	–	9	–	[24]
Ti	1,941	–	17	–	[24]
Co	1,768	–	14	–	[24]
Ru	2,607	–	13	–	[24]
Zn	693	–	17	–	[24]
Zr	2,125	–	13	–	[24]
Solid solution alloys					
Al-6 wt.% Mg	–	–	25	83	[2]
Al-7.5 wt.% Mg*	–	–	22	–	[19]

Al-7.5 wt.% Mg-0.3 wt.% Sc*	–	–	26	–	[19]
Al-6.8 wt.% Mg*	–	25	23	1,350 (HRTEM)	[18]
Fe-1.5 wt.% Mo	–	–	32	160	[28]
Fe-50 wt.% Co	–	–	12	3,800	[29]
Fe-40 wt.% Al	–	–	10	600	[30]
Cu-5 wt.% Nb	–	–	13	–	[20]
Cu-10 wt.% Nb	–	15	9	–	[20]
Cu-15 wt.% Nb	–	–	8	–	[20]
Cu-20 wt.% Nb	–	–	7	–	[20]
Cu-8 wt.% Mo	–	–	7	1,300	[31]
Dispersion alloys					
Fe-1.5 wt.% Mo-1.5 wt.% amorphous SiO_2	–	–	21	210	[28, 32]
Compounds					
SrF_2	–	–	30	70	[33]
CdF_2	–	–	28	77	[33]
BaF_2	–	–	27	82	[33]
CaF_2	–	–	25	52	[33]
WC	–	–	8	1,900	[34]
WC	–	–	10	–	[35]
$CoSb_3$	–	–	20	–	[36]

by SPD-processing of bulk materials. The high pressure induced by the collision between powder particles and balls during milling also retards dislocation annihilation as discussed in Chapter 2. This pressure has a similar value as applied on bulk materials by high-pressure torsion (about 6 GPa, see Chapter 2). The grain size obtained by TEM after milling is usually 10–20 nm. As below the grain size of ~20 nm, subgrains do not form inside the grains (see Chapter 5), therefore the crystallite and grain sizes agree well.

The smallest crystallite size values achieved in pure metal powders by milling at RT are plotted as a function of the melting point in Figure 4.6. If more than one crystallite size data are presented for a given material in Table 4.1, the smallest value is selected for Figure 4.1. Although there is no strict correlation between the melting point and the smallest

Figure 4.6 The smallest crystallite size in pure metal powders achieved by milling at RT as a function of the melting point. The data are taken from Table 4.1

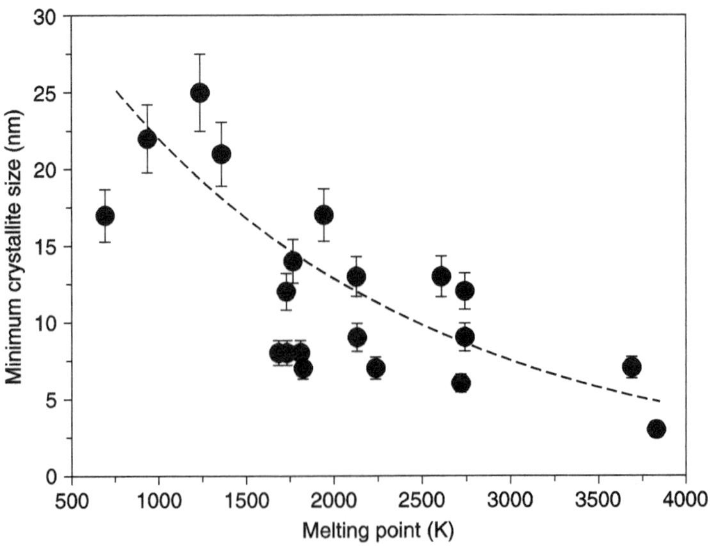

crystallite size, the metals having a higher melting point tend to exhibit smaller saturation crystallite size [23, 24]. For materials with higher melting points, RT corresponds to lower homologous temperature leading to weaker thermal activation for recovery and recrystallisation processes during milling. As a consequence, the maximum dislocation density and the minimum crystallite size are higher and smaller, respectively. As the annihilation processes of lattice defects are also influenced by other material parameters (e.g. shear modulus and stacking fault energy), the saturation dislocation density and crystallite size cannot be related solely to the melting point. It is interesting to note that the minimum grain size achieved by milling at RT is not reduced by decreasing the temperature to liquid nitrogen temperature (LNT), as shown in Table 4.1.

4.2 Defect structure in nanopowders produced by bottom-up approaches

One of the most frequently applied bottom-up processing routes of nanopowders is inert-gas condensation. Despite the lack of severe plastic deformation during this method, the as-processed nanoparticles contain dislocations with similar density, as in SPD-processed bulk ultrafine-grained (UFG) materials. For instance, in a Cu nanopowder produced by inert-gas condensation, the dislocation density was found to be 47×10^{14} m^{-2} [37], which is close to the value obtained by high-pressure torsion (see Chapter 2). However, these dislocations are most probably related to the particle growth mechanism as they have screw character as determined by x-ray line profile analysis, while the dislocations formed during SPD have edge character instead. The crystallite size determined by x-ray line profile analysis (~21 nm) agrees well with the grain size obtained by TEM (~16 nm), indicating the lack of subgrain structure inside grains. The average distance between dislocation is 15 nm as calculated from the inverse square-root of the dislocation density. Comparing this value with the grain size, it is suggested that a grain contains about one dislocation on average. Of course, the distribution of dislocations in the grain structure may be heterogeneous, i.e. the smaller grains may contain zero or only one dislocation while the larger grains may have more than two dislocations. Therefore, the relatively high dislocation density in inert-gas condensed Cu powder is caused only by the small grain size and not by the large number of dislocations inside the individual grains. Additionally, a significant value of twin boundary frequency, 3.6%, is detected by x-ray line profile analysis, which agrees well with the value determined by TEM (3.4%). TEM investigation shows that twin boundaries are distributed heterogeneously in the grain structure as they form preferably in the smaller grains.

A significant amount of twin boundaries was also detected in Pd nanoparticles processed by inert-gas condensation despite its large stacking fault energy (180 mJ/m^2) [38]. The twin boundary frequency was 2%, which corresponds to an average twin boundary spacing of 11 nm in Pd. The mean crystallite size, 18 nm, obtained by x-ray line profile analysis agrees well with the particle size, 19 nm, determined by TEM. Comparing the mean grain size and twin spacing values, the average number of twin boundaries in a grain is about two. The stacking fault frequency was practically zero in Pd nanoparticles produced by inert-gas condensation.

High dislocation density was found in nanocrystalline CeO_2 powder processed by calcination of grounded ceria xerogel [39]. The xerogel was produced by hydrolysis of a solution of cerium isopropoxide in isopropyl alcohol with the addition of acetylacetone. The calcination temperature varying between 300 and 600°C influences the grain size distribution and the dislocation density in ceria [40]. The mean crystallite size increases from 4 to 18 nm with increasing the calcination temperature from 300 to 600°C and the size distribution becomes much wider. The particle size determined by TEM is in good agreement with the crystallite size obtained by x-ray line profile analysis. The dislocation density is 110–140 × 10^{14} m^{-2} after calcination at temperatures between 300 and 500°C. At 600°C, the dislocation density is much lower, ~30 × 10^{14} m^{-2}. The character of dislocations is half edge, half screw.

Despite the high SFE of Al and Ni, their nanopowders often contain high density of twins. Figure 4.7a shows a TEM image of a Ni powder processed by electro-explosion of Ni wire [41]. The average particle size is ~100 nm as calculated from the particle size distribution presented in Figure 4.7b. The particles in Figure 4.7a have a significant amount of twin boundaries. The mean twin boundary

Figure 4.7 **(a) Bright-field TEM image of Ni powder processed by electro-explosion of Ni wire (b) the particle size distribution obtained from TEM images**

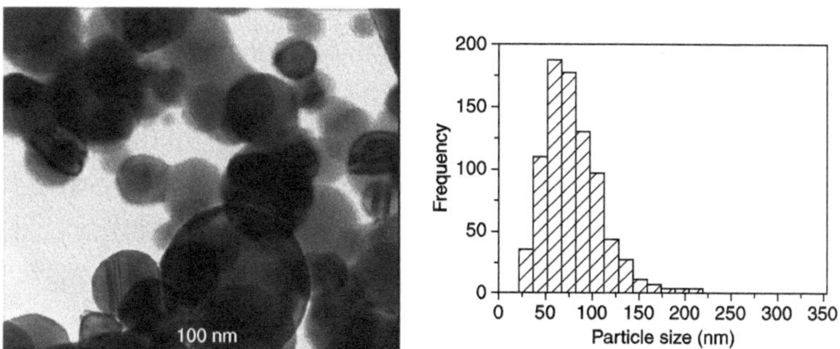

(a) (b)

Source: Reproduced from Gubicza et al (2007) [52] with permission from Cambridge University Press.

spacing is $d_{twin} \approx 6$ nm, which gives a twin boundary frequency of $\beta \approx 3\%$ using Equation 3.15 from Chapter 3. Twins are also observed at the neck of joined Al nanoparticles having a mean size of ~100 nm processed by cryomelting process. The neck may form by coalescence of crystalline nanoparticles due to their collision in the Ar flux transporting them into the powder collector. The high stresses related to the neck surface curvature may be relaxed by twin formation. Dislocations are not detected either in Ni processed by electro-explosion of wire or in cryomelted Al.

4.3 Effect of consolidation conditions on microstructure of sintered metals

During consolidation of nanopowders, significant changes usually occur in the grain structure as well as in the type and densities of lattice defects. Most sintering processes apply high pressure and temperature in order to minimise the remaining porosity in the consolidated materials. The application of high temperature during sintering may yield grain growth, therefore the consolidated materials may lose the unique properties that would be caused by their nanostructures. The temperature and/or duration of sintering can be reduced by the application of high pressure and/or current pulses in consolidation. Finer grain structure in sintered materials can also be achieved by smaller initial powder particles. The initial particle size of nanopowders and the consolidation conditions influence the grain size distribution and the lattice defect structure in the sintered materials. In the following, these effects are illustrated on Ni.

The Ni nanopowder selected for this study is processed by electro-explosion of Ni wires. The particle size distribution of the initial powder is presented in Figure 4.7a. The mean particle size is ~100 nm accordingly. It is revealed that the powder particles have spherical shape and the characteristic lattice defects inside the particles are twin faults (see Figure 4.7a). The twin boundary frequency estimated roughly from TEM images is about 3%. The energy-filtered TEM images in Figures 4.8a and 4.8b show the element maps for nickel and oxygen, respectively [41]. It can be seen that there is a

Figure 4.8 Energy-filtered TEM images on Ni nanopowder showing the element maps for (a) nickel and (b) oxygen, respectively. The lighter the colour in (a) and (b), the higher the local nickel and oxygen contents, respectively. The corresponding x-ray diffraction pattern is presented in (c) [41]

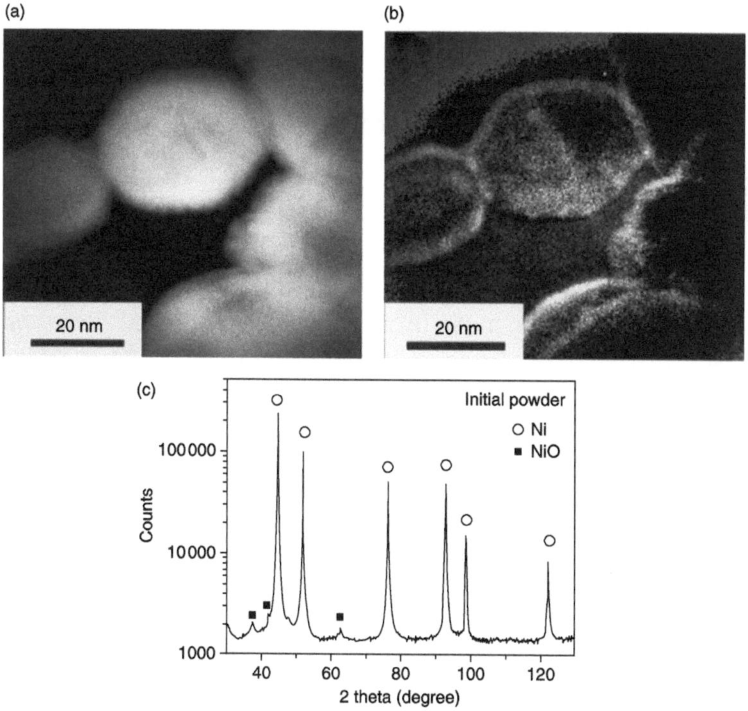

Source: Reproduced from Gubicza et al (2007) [52] with permission from Cambridge University Press.

significant amount of oxygen both in the surface layer (about 2 nm thick) and also inside the particles in the initial powder when it was exposed to air. However, the great majority of oxygen can be found on the surface of the particles. The x-ray diffraction pattern in Figure 4.8c shows that besides the main Ni phase NiO is also identified in the powder. The oxygen detected in the initial powder particles is most probably contained in the NiO phase. The powder is supplied in high vacuum-sealed glass capsules to avoid oxidation. The powder was consolidated by hot-isostatic pressing (HIP) and spark-plasma sintering (SPS) methods. Before HIP processing, the glass containing the powder was broken in a glove-box and subsequently encapsulated in a steel can and sealed under inert gas (Ar) to prevent oxidation. It takes about one week to remove gas introduced during the encapsulation step. During HIP processing, the capsule was subjected to a pressure of 140 MPa at 700°C for 150 minutes. As for the SPS processing, the capsule containing the powder was broken in air and rapidly transferred to a graphite mould. During SPS procedure, the powder was held under a pressure of 150 MPa at 500°C for 1 minute while pulses of high current density (of the order of 1,000 A/cm^2) were applied to the sample for promoting consolidation. The most important advantage of the latter method is that the compaction occurs more quickly and at lower temperatures than in the case of classical techniques (e.g. HIP), thus preventing grain growth. The densities relative to the value of bulk pure Ni are 95.5 ± 0.1 and $95.4 \pm 0.1\%$ for the HIP and SPS-processed samples, respectively [41].

For comparison of the NiO phase content in the initial powder and the consolidated samples, the integrated intensity ratio of the NiO and Ni peaks on the x-ray diffractograms at $2\Theta = 37.4°$ and $44.6°$, respectively, is determined. It should be noted that this ratio does not give the NiO phase content in the samples; it is only proportional with the volume fraction of the NiO phase. The intensity ratio for the initial powder was found to be 0.5%, which does not change during HIP-processing as a result of the careful isolation of the sample from air. On the other hand, SPS procedure results in a material having significantly higher oxide content (the intensity ratio is 0.9%), which can be attributed to the powder processing in air before consolidation [41].

Figures 4.9a and 4.9b show TEM images obtained on the HIP and SPS-processed samples, respectively. The grain size distributions determined from TEM are presented in

Figure 4.9 TEM images of the sample processed by (a) HIP and (b) SPS. Grain-size distributions for the (c) HIP- and (d) SPS-processed bulk samples determined from TEM images. The distributions of twinned grains are also shown by hatched columns [41]

Source: Reproduced from Gubicza et al (2007) [52] with permission from Cambridge University Press.

Figures 4.9c and 4.9d. The mean grain sizes for the HIP- and SPS-processed samples were 403 and 294 nm, respectively. The mean grain size values for the consolidated samples are four to five times larger than the mean particle size of the initial powder, which indicates that the high temperature consolidation results in a grain coarsening. The smaller grain size in the specimen produced by SPS is attributed to the lower temperature and the shorter duration of processing, which reduced the grain-growth during consolidation. The mean crystallite size, the dislocation density and the twin boundary frequency for the consolidated materials are listed in Table 4.2. TEM and x-ray line profile analysis show that the twin boundary frequency decreases from ~3% to ~0.3% during sintering either by HIP or SPS method (see Table 4.2). At the same time, while dislocations were not observed in the initial powder by TEM, a significant amount of dislocation was formed ($5–6 \times 10^{14}$ m^{-2}) during consolidation [41]. The increase of the dislocation density during sintering indicates that the dislocation controlled plastic deformation has a significant contribution to consolidation. However, the values of the dislocation density in the consolidated samples are much smaller than in the bulk UFG materials processed by SPD procedures (see Chapter 2). The annihilation of growth twins during consolidation may occur by grain boundary or twin fault migration during grain growth [42], as twin boundaries have high excess energies due to the large stacking fault energy in Ni. Additionally, the interaction between dislocations and twins may also contribute to the reduction of twin boundary frequency (untwinning), as discussed in Chapter 3. Although formation of twin boundaries during consolidation is not excluded, the stronger annihilation processes yield a net reduction of twin boundary frequency. Figures 4.9c and 4.9d show the distribution of grains containing twins in the samples produced by HIP and SPS,

Table 4.2 The intensity ratio (I_{NiO}/I_{Ni}) of the x-ray peaks for NiO and Ni at $2\Theta = 37.4°$ and $44.6°$, respectively, the mean grain size obtained from TEM images, the mean crystallite size, the dislocation density and the twin boundary frequency for the samples processed by HIP and SPS [41]

Powder particle size [nm]	Consolidation method	I_{NiO}/I_{Ni} [%]	Grain size [nm]	Crystallite size [nm]	Dislocation density [10^{14} m^{-2}]	Twin boundary frequency [%]
100	HIP	0.5±0.1	403	140±15	5.6±0.5	0.32±0.03
100	SPS	0.9±0.1	294	153±16	5.1±0.5	0.29±0.03
50	SPS	1.5±0.1	250	140±14	3.7±0.4	0.25±0.03

respectively. The relative fractions of twinned grains are 19% and 17% in the HIP- and SPS-processed specimens, respectively [41]. The size distributions of twinned grains follow the total grain size distributions for both samples. The crystallite size is about two to three times smaller than the grain size, similarly as in the case of SPD-processed bulk UFG metallic materials (see Chapter 2). The difference can be explained by the fact that the crystallite size corresponds rather to the subgrain size.

In order to investigate the effect of the initial powder particle size on the defect structure, an additional material is consolidated by SPS from a powder having a mean particle size of 50 nm. The integrated intensity ratio of the NiO and Ni peaks on the x-ray diffractogram for this powder (1.2%) is about two times higher than for the initial powder with the particle size of 100 nm (see Table 4.2). The higher volume fraction of NiO in the powder having smaller particle size can be attributed to its larger specific surface area, where the great majority of the oxygen can be found (see Figure 4.8). The smaller initial powder particle size results in a smaller grain size after SPS; however, it has only a slight influence on the defect structure (crystallite size, dislocation density and twin boundary frequency) inside the grains under the present processing conditions, as revealed by Table 4.2. The mechanical behaviours of Ni samples consolidated from different nanopowders and under various conditions are compared in Chapter 5. It is noted that the grain growth during sintering can also be lowered by the addition of nanodisperse secondary phase (e.g. ceramic) powder particles. For instance, the crystallite size of ball milled Fe-1.5 wt.% Mo powder is 32 nm, which increases to ~255 nm during sintering by SPS at 800°C. When 1.5 wt.% SiO_2 is added to the powder before milling, the crystallite size in the SPS-processed sample is only 112 nm due to the pinning effect of SiO_2 particles [28]. The crystallite size obtained by x-ray line profile analysis shows a good agreement with the grain size determined by TEM. Moreover, the SPS-processed Fe-1.5 wt.% Mo-1.5 wt.% SiO_2 material has similar high dislocation density, ~10^{16} m^{-2}, than that measured immediately after ball milling of the initial powder. In the case of ball milled Cu powder, the short duration (2 min) and the low temperature (350°C that corresponds to the homologous temperature of 0.46) of SPS-processing result in a very small crystallite size of 32 nm, which agrees well with the grain size, 26 nm, as determined by TEM [43]. The dislocation density in this material, 16×10^{14} m^{-2}, is similar to the saturation value in Cu achievable by equal-channel angular pressing (ECAP) at RT. Experiments reveal that grain growth in consolidation can be avoided by applying high-pressure torsion (HPT) under very high pressure (~6 GPa) at RT. For instance, the grain size of cryomilled Al – 7.5 wt.% Mg powder, 22 nm, remained unchanged during consolidation by HPT [44].

4.4 Defect structure in metals sintered from blends of powders with different particle sizes

UFG materials usually exhibit higher tensile strength but lower ductility in comparison with their coarse-grained counterparts. The low ductility has been attributed to the absence of work hardening that leads to plastic instability (see Chapter 5). One strategy for the improvement of ductility of UFG materials is the incorporation of coarse grains in a fine-grained matrix [45–47]. In this case, the improvement of ductility can be attributed to the larger dislocation activity of coarse grains (storage and mobility), thereby enhancing the strain hardening capability of materials. In the following, the effect of addition of coarse particles to nanopowders on the consolidated microstructure is shown. Bulk samples were processed from blends of high purity (~99.9%) Ni nanopowder (NP) having an average particle size of ~100 nm and a conventional coarse-grained powder (CP) [48]. The particles in powder CP are agglomerated into particle clusters of ~50 μm in diameter as is revealed by the scanning electron microscopy (SEM) image in Figure 4.10a. The average size of the individual particles within the clusters was ~5 μm (see Figure 4.10b).

Along with the bulk samples processed from 100% NP and 100% CP powders (hereafter referred to as samples UFG and CG, respectively), two additional specimens with different volume fractions of powders NP and CP were prepared. Sample A was processed by mixing ~33 vol.% CP and ~67 vol.% NP powders while for sample B the relative amount of powder CP was only ~17 vol.% [48]. To reach an optimum distribution of the CP particles within powder NP, the powders were mixed in a three-dimensional shaker mixer for 6 h at RT. After degassing at RT for one week, the specimens were consolidated by HIP. The processing conditions were

Figure 4.10 SEM picture showing clusters in powder CP with an individual particle size of about 5 μm

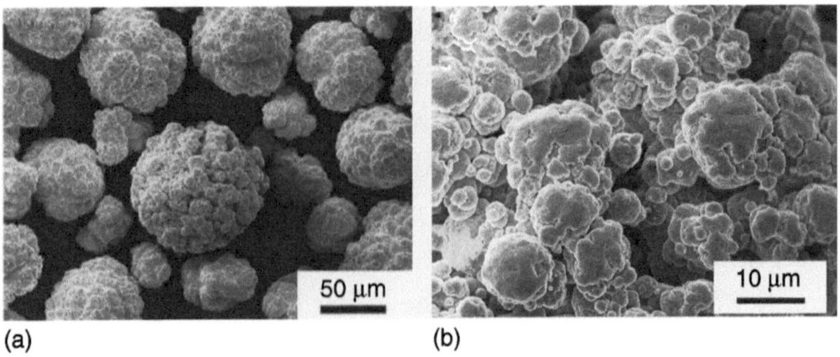

Source: Reprinted from Dirras et al. (2010) [48] with permission from Elsevier.

optimised to reach the desired microstructure having small porosity. The pressure, the temperature and the duration of HIP-processing were set around 150 MPa, 700°C and 150 minutes, respectively, except in the case of sample B, for which the temperature and duration were 640°C and 60 minutes, respectively. It is noted that the microstructure of sample UFG has already been shown in the previous section.

After HIP-processing, the relative densities are 98%, 97%, 94% and 96% for specimens CG, A, B and UFG, respectively. In sample CG, no significant amount of NiO phase is detected by x-ray diffraction. The higher the fraction of powder NP in the blend, the higher the NiO content in the consolidated specimen, since powder NP contains a considerable amount of NiO, as shown in the previous section [48]. Figure 4.11 shows SEM images of the microstructure for samples CG, A and B. These SEM images suggest that for the sake of simplicity, the composite microstructures in samples A and B can be considered as being bimodal grain structures consisting of coarse-grained (CG) and ultrafine-grained (UFG) components having grains with sizes smaller and larger than 1 μm, respectively. The fractions of components CG and UFG are 61% and 39% for sample A, respectively, while these values are 39% and 61% for sample B. For both bulk samples, the fraction of component CG evaluated from SEM pictures is higher than the volume fraction of powder CP in the powder blends measured before consolidation (33% and 17% for samples A and B, respectively). This difference can be explained by the denser particle packing in powder CP compared with powder NP, as supported by the agglomeration of the particles in Figure 4.10. The grain size in sample CG is 50–100 μm, which is similar or larger than the particle cluster size in powder CP, indicating a coalescence of powder particles during sintering, which may also spread into the neighbouring agglomerates. In samples A and B, the CG

Figure 4.11 **SEM images of the microstructure for samples CG (a), A (b) and B (c)**

Source: Reprinted from Dirras et al. (2010) [48] with permission from Elsevier.

clusters are homogeneously distributed within the UFG matrix (see Figures 4.11b and 4.11c). The average size of the CG clusters was the same (about 50 µm) as in the initial CP powder, i.e. the CP agglomerate size remained unchanged during consolidation of the blended specimens [48]. At the same time, the mean grain size inside the CG clusters decreases with increasing UFG fraction, and their values are 13 µm and 7 µm for samples A and B, respectively. In the composite samples it seems that the UFG component acts against the coalescence of CG grains as the higher the UFG content, the lower the level of grain growth in the CG component. Most probably, during powder mixing, NP particles got in between CG agglomerates, which formed thin intergranular UFG domains between CG grains in the consolidated samples. The higher oxide content in this UFG component hinders the grain-boundary motion of CG grains during sintering, thus lowering the grain-coalescence. This phenomenon is referred to as *barrier effect* of UFG fraction on coalescence of CG grains. Additionally, the lower degree of grain-coalescence in component CG in sample B can be partially attributed to the lower temperature and shorter time of sintering. It is noted that thermodynamic consideration suggests that coarser grains should grow at the expense of UFG grains during sintering of powder blends due to the higher specific grain boundary area of the smaller grains. Consequently, if the UFG fraction had been oxide-free, its addition to the CG component most probably would have accelerated the grain coarsening during sintering. According to TEM experiments, the average grain size of UFG grains in samples UFG, A and B is about 400 nm irrespective of the fraction of component CG. EBSD investigations show that in both samples A and B, the majority of the boundaries (92% and 95% for samples A and B, respectively) are high-angle grain boundaries [48].

The defect structure is determined by x-ray line profile analysis only for the component UFG in samples A and B, as the diffraction peaks reflected from the CG grains are as narrow as the instrumental profiles; therefore these lines cannot be evaluated for the microstructure. The dislocation densities in the UFG volumes of samples A and B are $0.7 \pm 0.2 \times 10^{14}$ m^{-2} and $0.3 \pm 0.2 \times 10^{14}$ m^{-2}, respectively, which are much lower than in sample UFG [48]. The twin boundary frequency is under the detection limit of x-ray line profile analysis in components UFG of samples A and B. The much lower defect densities in component UFG in the blended samples can be explained by a *shielding effect* of component CG on the UFG fraction during consolidation. The *shielding effect* means that when the pores are filled with the help of plastic deformation of particles in the blended materials, the softer CP particles deform preferably instead of the harder NP particles. The NP particles are harder due to their higher oxide content. Moreover, the NP particles in the pockets of CP clusters are most probably protected from the external stresses. As a consequence of the moderate deformation of NP particles, the dislocation density and the twin boundary frequency are smaller in component UFG of samples A and B than in sample UFG [48]. The strong reduction of twin boundary frequency in component UFG during consolidation of samples A and B is attributed to the annihilation of growth twins as a result of the high excess energy of twin boundaries in Ni having large SFE. Despite the lower CG content of sample B, which should result in a weaker *shielding effect*, the dislocation density and the twin boundary frequency in the UFG component of sample B are not higher than the values obtained for sample A. This may be attributed to the shorter sintering time in the case of sample B. The influence of CG fraction on the mechanical behaviour of the composite materials is presented in Chapter 5.

4.5 Evolution of microstructure during consolidation of diamond and ceramic nanopowders

The application of high pressure during sintering of metallic nanopowders facilitates the consolidation as the shear stresses induced at the contact points of particles yield plastic deformation, which contributes to the filling in of the pores between particles. Experiments reveal that plastic deformation also occurs during sintering of diamond or ceramic powders at high pressures and homologous temperatures of several GPa and ~0.4, respectively [49, 50, 51]. If the initial powder is coarse-grained, the plastic deformation during consolidation yields a decrease in grain size and an increase in dislocation density. For instance, the initial grain size of 30–40 μm in diamond powder is refined to ~45 nm during sintering at 2 GPa and 1,470 K, while dislocations with the density of 12×10^{14} m^{-2} are formed [49]. When the particles in the initial diamond powder are nanosized, such grain-refinement does not occur, although the dislocation density increases. This is a consequence of the low dislocation storage capability of very small crystallites irrespective of their material. As an example, Figure 4.12 shows the crystallite size and the dislocation density in bulk diamonds consolidated from a nanopowder with an average

Figure 4.12 Dislocation density and the crystallite size in nanodiamond samples as a function of consolidation pressure

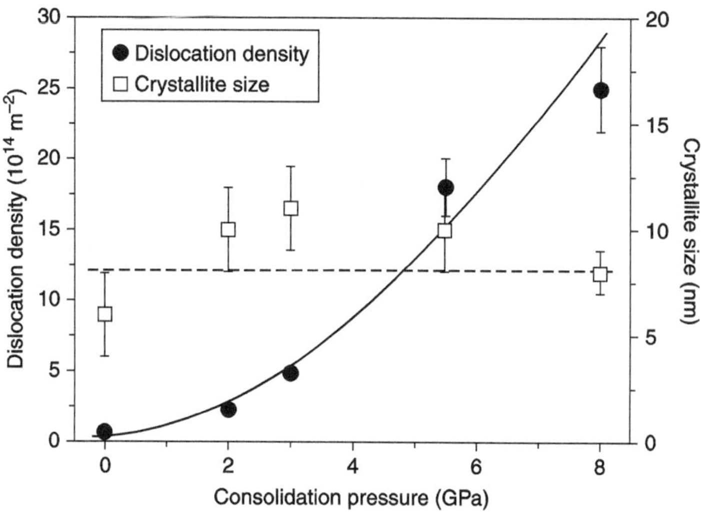

particle size of 10 nm at different pressures and 1,470 K for 60 seconds [50]. The mean crystallite size in the initial powder determined by x-ray line profile analysis has similar value (6 ± 2 nm) as the particle size obtained by TEM (10 nm). The crystallite size remains nearly unchanged during sintering, while the dislocation density increases with increasing pressure, since the higher pressure is accompanied by larger shear stresses at the contact points of particles.

The dislocation activity is also found during sintering of nanocrystalline α-Si_3N_4 ceramics by shock-compaction (shock-wave consolidation) [51]. Experiments prove that the consolidation is achieved through both densification by plastic deformation and interparticle bonding by surface melting of particles induced by the shock loading. The mean particle size of the initial powder is 70 nm. If the shock pressure and temperature are under 40 GPa and 1,200 K, respectively, the plastic deformation yields an increase of the dislocation density and a decrease of the grain size to 40 nm, most probably due to fragmentation of the initial particles into subgrains. At higher pressures and temperatures, the melting propagates from the surface into the grain interior of α-Si_3N_4 particles, and β-Si_3N_4 grains with the size of 200–500 nm grow from the molten mass. The hardness of the consolidated ceramics is optimal before the occurrence of this transformation [51].

The influence of sintering pressure and temperature on the defect structure in low SFE ceramics is illustrated with the case of SiC materials consolidated from a nanopowder with a particle size of 30 nm [52]. The sintering procedure was carried out at temperatures of 1,400, 1,600 and 1,800°C. At each temperature, specimens were sintered at pressures of 2, 4 and 5.5 GPa. This process gave nine bulk sintered SiC samples, in addition to the initial powder. An additional specimen was produced at 1800°C and at a very high sintering pressure of 8 GPa. In each case the sintering time was 10 seconds.

X-ray line profile analysis carried out for the initial powder gives the mean crystallite size of 8.3 ± 0.7 nm and the twin boundary frequency of 7.4 ± 0.6%, while the dislocation density is below the limit of detection by x-ray line broadening, i.e. $\rho < 10^{13}$ m^{-2}. From the results it can be concluded that in the initial SiC nanopowder, the main lattice defects are planar faults, in particular twin boundaries most probably due to the extremely low SFE of SiC (0.2 mJ/m^2 [53]). Table 4.3 shows the microstructure parameters of nine

Table 4.3 **The parameters of the microstructure for nanocrystalline SiC specimens sintered at different pressures and temperatures. $\langle x \rangle_{area}$ is the area-weighted mean crystallite size, β is the twin boundary frequency and ρ is the dislocation density. The samples containing a significant amount of dislocations are marked by light grey colour [52]**

Temperature [°C]	Pressure [GPa]		
	2	4	5.5
1,400	$\langle x \rangle_{area}$ = 8.9±0.8 nm β=9.4±0.9% $\rho < 10^{13}$ m^{-2}	$\langle x \rangle_{area}$ = 15±1 nm β=11±1% $\rho < 10^{13}$ m^{-2}	$\langle x \rangle_{area}$ = 19±1 nm β = 11±1% $\rho < 10^{13}$ m^{-2}
1,600	$\langle x \rangle_{area}$ =13±1 nm β = 11±1% $\rho < 10^{13}$ m^{-2}	$\langle x \rangle_{area}$ =16±1 nm β = 11±1% $\rho < 10^{13}$ m^{-2}	$\langle x \rangle_{area}$ =94±7 nm β = 0.20±0.03% ρ = 4.0±0.8 × 10^{14} m^{-2}
1,800	$\langle x \rangle_{area}$ = 13±1 nm β = 11±1% $\rho < 10^{13}$ m^{-2}	$\langle x \rangle_{area}$ = 124±10 nm β=0.10±0.02% ρ = 3.0±0.8 × 10^{14} m^{-2}	$\langle x \rangle_{area}$ = 126±10 nm β = 0.17±0.03% ρ = 4.0±0.8 × 10^{14} m^{-2}

Source: Reproduced from Gubicza et al (2007) [52] with permission of Cambridge University Press.

nanocrystalline SiC specimens sintered at pressures not larger than 5.5 GPa. The crystallite size increases as a result of sintering. The crystallite size increases with the temperature for a given pressure and also with the pressure at a constant temperature. The crystallite size may increase due to coalescence of grains, similar to Ostwald-ripening. The increase of temperature at constant pressure causes the increase of mobility of atoms, resulting in the faster grain growth. Although the higher pressure usually reduces the mobility of atoms, the increased elastic shear strains at the contact surfaces may increase the driving force for reducing the fraction of grain boundaries, i.e. induce grain growth.

At low pressure values, twins are formed during sintering and the dislocation density remains below the detection limit of line profile analysis. At 1,600°C and 1,800°C, and above a certain pressure limit, the planar fault density decreases to a very low level, i.e. to about 0.1±0.02%, and dislocations become the main type of lattice defects. The values of the pressure limit are 5.5 and 4 GPa at 1,600°C and 1,800°C, respectively, i.e. this limit decreases with increasing temperature [52]. The decrease of the planar fault density when the pressure is increasing from 2 to 5.5 GPa at 1,800°C is also justified by the comparison of TEM images in Figures 4.13a and 4.13b. The values of twin boundary spacing determined from the TEM images are 3–6 nm and 100–200 nm for the specimens sintered at 2 and 5.5, respectively. Taking into account that the distance between the neighbouring {111} planes in SiC is d_{111} = 0.252 nm and using Equation 3.15, the twin boundary frequency is obtained to be in the range of 6–12% and 0.1–0.2% for the samples prepared at 2 and 5.5 GPa, respectively. These values are close to the twin boundary frequencies determined by x-ray line profile analysis (see Table 4.3).

For the specimen prepared at 8 GPa and 1,800°C the crystallite size, the twin boundary frequency and the dislocation

Figure 4.13 (a) TEM images of specimens sintered at 1,800 °C and 2GPa, (b) 5.5GPa, and (c) 8GPa. Please note the different magnification for (b) [52]

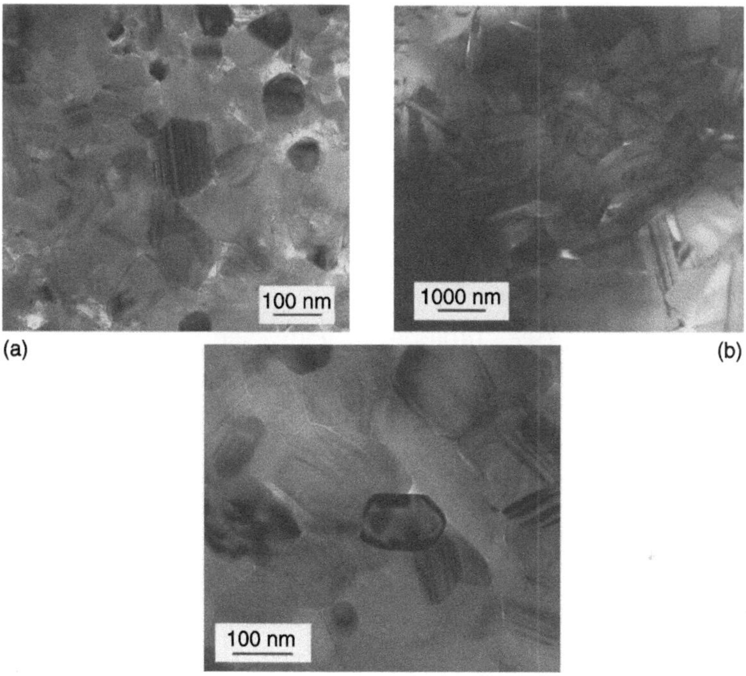

(a) (b) (c)

Source: Reproduced from Gubicza et al (2007) [52] with permission of Cambridge University Press.

density are 73±8 nm, 1.8±0.3% and 15±2 × 10^{14} m^{-2}, respectively [52]. The crystallite size decreased while both the dislocation density and the twin boundary frequency increased compared with the sample processed at 5.5 GPa at the same temperature. The TEM images shown in Figure 4.13b and 4.13c confirm the higher planar fault density and the smaller grain size at 8 GPa compared with the specimen sintered at 5.5 GPa. Although the higher shear strains at 8 GPa would induce higher driving force for the increase of the grain size, the diffusion rate and rearrangement of atoms are impeded with the increased pressure and these effects could slow down grain growth when pressure exceeds a critical value. The distance between the twin boundaries in the TEM images is 10–20 nm for the specimen sintered at 8 GPa, which gives 1–2% for the twin boundary frequency, which is close to the value determined by x-ray line profile analysis (see Table 4.3). The values of the grain sizes observed in the TEM images of Figure 4.13 are 30–100 nm, 100–300 nm and 80–120 nm for the specimens sintered at 1,800°C and 2, 5.5 and 8 GPa, respectively [52]. The larger grain sizes for 5.5 and 8 GPa than for 2 GPa are consistent with the trend observed for the crystallite size (see Table 4.3). However, the grain sizes obtained by TEM are somewhat larger than the crystallite sizes obtained by x-ray line profile analysis. This difference can be attributed to the fact that x-ray line profile analysis measures the coherently scattering domain size, which is generally smaller than the grain size observable by microscopic methods. The coherency of x-rays scattered from a grain can be broken down by special arrangement of dislocations, such as dipolar dislocation walls or low angle boundaries, which gives no or only weak contrast difference in TEM images [54].

The crystallite size, the dislocation density and the twin boundary frequency for the 10 sintered specimens as a function of the sintering pressure and temperature are plotted in Figures 4.14a, 4.14b and 4.14c, respectively. It seems that at high temperatures and pressures the relatively large crystallite size enables the formation of dislocations during the sintering process. The elimination of planar faults when grain growth takes place during firing at high temperatures has already been observed for SiC [55, 56]. This result is also in line with observations for fcc metals, where planar faults and dislocations control plasticity below and above a critical crystallite size, respectively [57]. The value of the critical

Figure 4.14 **(a) The crystallite size, (b) the dislocation density and (c) the twin density for 10 sintered specimens as a function of the sintering pressure and temperature. The wire grids in the figures are to guide the eye [52]**

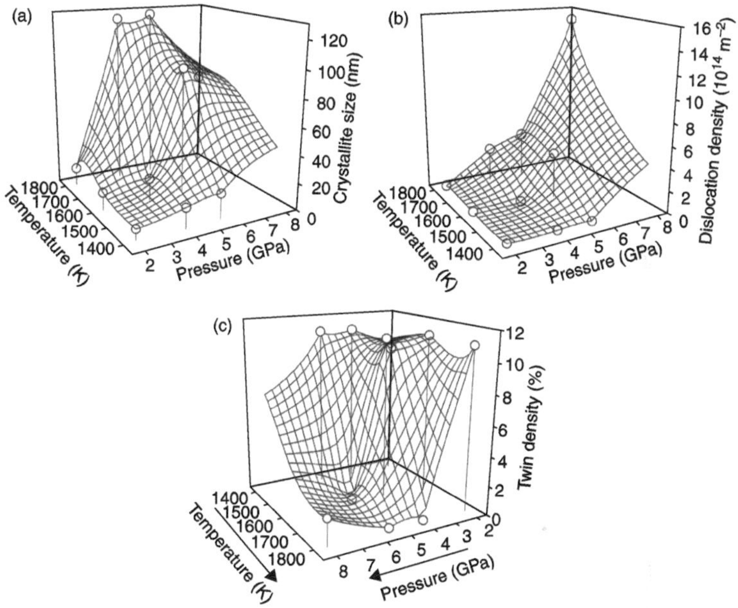

Source: Reproduced from Gubicza et al (2007) [52] with permission of Cambridge University Press.

crystallite size where that transition takes place depends on the properties of materials. For example, for Cu ~40 nm is suggested as the critical crystallite size, below which deformation proceeds by twinning rather than by dislocation glide [57, 58]. The data presented in Table 4.3 do not allow us to precisely identify the critical grain size for which dislocations become prevalent, but it appears to be bounded between 20 and 90 nm. It has been mentioned above that the pressure limit, where dislocations become the main lattice defects, decreases with increasing temperature. This phenomenon can be explained by the promoting effect of temperature for the growth of crystallites [59]. It is interesting to note that when the crystallite size decreases, e.g. for sintering at 8 GPa and 1,800°C, the planar defects have significant densities, which again supports the observed correlation between the crystallite size and the defect structure.

It is noted that the grain growth during consolidation of SiC powders at high temperature can be suppressed by the application of SPS method [59]. For instance, SPS-processing of SiC nanopowder with the crystallite size of about 10 nm at 1,700°C for 10 minutes under a pressure of 40 MPa yields only a moderate growth of crystallites to 25 nm [59]. The TEM analysis gives a slightly larger grain size of 40 nm. Despite the small grain size, the stacking disorder is lowered during sintering, which indicates the role of high temperature in the decrease of planar faults in nanocrystalline SiC.

It is also noted that if the dislocation density is very high in the initial powder, it can decrease simultaneously with the grain-growth during high-temperature sintering. It is manifested during consolidation of WC nanopowder processed by high-energy ball-milling. The crystallite size and the dislocation density in the ball-milled powder are 8 nm and 1.9×10^{17} m^{-2}, respectively. After sintering at 1,420°C and 7 MPa, the crystallite size increases to 113 nm, while the dislocation density decreases to 2×10^{15} m^{-2} [60].

4.6 References

1. R.A. Varin, T. Czujko and Z.S. Wronski, *Nanomaterials for Solid State Hydrogen Storage*, Springer Science+Business Media, LLC 2009, ISBN 978-0-387-77711-5.
2. J. Gubicza, M. Kassem, G. Ribárik and T. Ungár, 'The microstructure of mechanically alloyed Al–Mg

determined by X-ray diffraction peak profile analysis', *Materials Science and Engineering*, A 372 (2004) 115–122.

3. J.L. Murray, 'The Al–Mg (Aluminum–Magnesium) system', *Bulletin of Alloy Phase Diagrams*, 3 (1982) 60–74.
4. M. Hansen and K. Anderko, *Constitution of Binary Alloys*, Met. Eng. Series, Eds. R.F. Mehl and M.B. Bever, McGraw-Hill, N.Y., 1958, pp. 106–107.
5. C. Suryanarayana, 'Mechanical alloying and milling', *Progress in Materials Science*, 46 (2001) 1–184.
6. Á. Révész, J. Lendvai and T. Ungár, 'Melting Point Depression and Microstructure in Ball-Milled Nanocrystalline Aluminium Powders', *Materials Science Forum*, 343–346 (2000) 326–331.
7. T. Ungár, I. Dragomir, Á. Révész and A. Borbély, 'The contrast factors of dislocations in cubic crystals: the dislocation model of strain anisotropy in practice', *Journal of Applied Crystallography*, 32 (1999) 992–1002.
8. I. Kovács and J. Lendvai, 'The mechanism of clustering in supersaturated solid solutions of A1-Mg2Si alloys', *E. Nagy, Acta Metallurgica*, 20 (1972) 975–983.
9. D.A. Hughes, 'Microstructural evolution in a non-cell forming metal: Al–Mg', *Acta Metallurgica et Materialia*, 41 (1993) 1421–1430.
10. G.F. Dirras, M.-P. Biget and C. Rey, 'On the microstructural evolution of cold-rolled Al+ 5at.% Mg', *Scripta Metallurgica et Materialia*, 33 (1995) 755–760.
11. A. Molinari, I. Lonardelli, K. Demetrio and C. Menapace, 'Effect of the particle size on the thermal stability of nanostructured aluminum powder: dislocation density and second-phase particles controlling the grain growth', *Journal of Materials Science*, 45 (2010) 6739–6746.

12. S. Vives, E. Gaffet and C. Meunier, 'X-ray diffraction line profile analysis of iron ball milled powders', *Materials Science and Engineering*, A 366 (2004) 229–238.
13. O. Boytsov, A.I. Ustinov, E. Gaffet and F. Bernard, 'Correlation between milling parameters and microstructure characteristics of nanocrystalline copper powder prepared via a high energy planetary ball mill', *Journal of Alloys and Compounds*, 432 (2007) 103–110.
14. M. Abdellaoui and E. Gaffet, 'The physics of mechanical alloying in a modified horizontal rod mill: mathematical treatment', *Acta Materialia*, 43 (1996) 1087–1098.
15. R.K. Khatirkar and B.S. Murty, 'Structural changes in iron powder during ball milling', *Materials Chemistry and Physics*, 123 (2010) 247–253.
16. A.H.W. Ngan and M. Wen, 'Atomistic simulations of Paidar–Pope–Vitek lock formation in Ni_3Al', *Computational Materials Science*, 23 (2002) 139–145.
17. H. Van Swygenhoven, A. Caro and D. Farkas, 'A molecular dynamics study of polycrystalline fcc metals at the nanoscale: grain boundary structure and its influence on plastic deformation', *Materials Science and Engineering*, A 309–310 (2001) 440–444.
18. F. Zhou, X.Z. Liao, Y.T. Zhu, S. Dallek and E.J. Lavernia, 'Microstructural evolution during recovery and recrystallization of a nanocrystalline Al-Mg alloy prepared by cryogenic ball milling', *Acta Materialia*, 51 (2003) 2777–2791.
19. F. Zhou, D. Witkin, S.R. Nutt and E.J. Lavernia, 'Formation of nanostructure in Al produced by a low-energy ball milling at cryogenic temperature', *Materials Science and Engineering*, A 375–377 (2004) 917–921.

20. R.S. Lei, M.P. Wang, Z. Li, H.G. Wei, W.C. Yang, Y.L. Jia and S. Gong, 'Structure evolution and solid solubility extension of copper-niobium powders during mechanical alloying', *Materials Science and Engineering*, A 528 (2011) 4475–4481.
21. M. Samadi Khoshkhoo, S. Scudino, J. Thomas, K.B. Surreddi and J. Eckert, 'Grain and crystallite size evaluation of cryomilled pure copper', *Journal of Alloys and Compounds*, 509S (2011) S343–S347.
22. C.C. Koch, 'Synthesis of nanostructured materials by mechanical milling: problems and opportunities', *Nanostructured Materials*, 9 (1997) 13–22.
23. J. Eckert, J.C. Holzer, C.E. Krill and W.L. Johnson, 'Structural and thermodynamic properties of nanocrystalline fcc metals prepared by mechanical attrition', *Journal of Materials Research*, 7 (1992) 1751–1761.
24. F.A. Mohamed and Y. Xun, 'Correlations between the minimum grain size produced by milling and material parameters', *Materials Science and Engineering*, A 354 (2003) 133–139.
25. Á. Révész, 'Melting behavior and origin of strain in ball-milled nanocrystalline Al powders', *Journal of Materials Science*, 40 (2005) 1643–1646.
26. P. Scardi and M. Leoni, 'Whole powder pattern modelling', *Acta Crystallographica*, A 58 (2002) 190–200.
27. Á. Révész, T. Ungár, A. Borbély and J. Lendvai, 'Dislocations and grain size in ball-milled iron powder', *Nanostructured Materials*, 7 (1996) 779–788.
28. S. Libardi, M. Zadra, F. Casari and A. Molinari, 'Mechanical properties of nanostructured and ultrafine-grained iron alloys produced by spark plasma sintering of ball milled powders', *Materials Science and Engineering*, A 478 (2008) 243–250.

29. H. Moumeni, A. Nemamcha, S. Alleg and J.M. Greneche, 'Stacking faults and structure analysis of ball-milled Fe–50% Co powders', *Materials Chemistry and Physics*, 122 (2010) 439–443.
30. M. Mhadhbi, M. Khitouni, L. Escoda and J.J. Sunol, 'X-ray studies of structure defects in nanostructured FeAl alloy', *Materials Letters*, 64 (2010) 1802–1805.
31. C. Aguilar, D. Guzman, P.A. Rojas, S. Ordonez and R. Rios, 'Simple thermodynamic model of the extension of solid solution of Cu-Mo alloys processed by mechanical alloying', *Materials Chemistry and Physics*, 128 (2011) 539–542.
32. S. Libardi, M. Leoni, L. Facchini, M. D'Incau, P. Scardi and A. Molinari, 'Effect of the dispersion of nanometric silica particles on the thermal stability of a nanostructured iron based powder', *Materials Science and Engineering*, A 445–446 (2007) 244–250.
33. G. Ribárik, N. Audebrand, H. Palancher, T. Ungár and D. Louer, 'Dislocation densities and crystallite size distributions in nanocrystalline ball-milled fluorides, MF2 (M= Ca, Sr, Ba and Cd), determined by X-ray diffraction line-profile analysis', *Journal of Applied Crystallography*, 38 (2005) 912–926.
34. T. Ungár, A. Borbély, G.R. Goren-Muginstein, S. Berger and A.R. Rosen, 'Particle-size, size distribution and dislocations in nanocrystalline tungsten-carbide', *Nanostructured Materials*, 11 (1999) 103–113.
35. A.S. Kurlov and A.I. Gusev, 'Model for milling of powders', *Technical Physics*, 56 (2011) 975–980.
36. L. Zhang, A. Grytsiv, M. Kerber, P. Rogl, E. Bauer, M.J. Zehetbauer, J. Wosik and G.E. Nauer, 'MmFe4Sb12- and CoSb3-based nano-skutterudites prepared by ball milling: Kinetics of formation and transport properties', *Journal of Alloys and Compounds*, 481 (2009) 106–115.

37. T. Ungár, S. Ott, P.G. Sanders, A. Borbély and J.R. Weertman, 'Dislocations, grain size and planar faults in nanostructured copper determined by high resolution X-ray diffraction and a new procedure of peak profile analysis', *Acta Materialia*, 46 (1998) 3693–3699.
38. C.E. Krill and R. Birringer, 'Estimating grain-size distributions in nanocrystalline materials from X-ray diffraction profile analysis', *Philosophical Magazine*, 77 (1998) 621–640.
39. P. Scardi, M. Leoni, M. Müller and R. Di Maggio, 'In situ size-strain analysis of nanocrystalline ceria growth', *Materials Science and Engineering*, A 528 (2010) 77–82.
40. M. Leoni and P. Scardi 'Whole Powder Pattern Modelling: Theory and Applications', in: E.J. Mittemeijer and P. Scardi (eds), *Diffraction Analysis of the Microstructure of Materials*, Springer Verlag, Berlin (2004), pp. 413–454.
41. J. Gubicza, H-Q. Bui, F. Fellah and G.F. Dirras, 'Microstructure and mechanical behavior of ultrafine-grained Ni processed by different powder metallurgy methods', *Journal of Materials Research*, 24 (2009) 217–226.
42. J. Wang, N. Li, O. Anderoglu, X. Zhang, A. Misra, J.Y. Huang and J.P. Hirth, 'Detwinning mechanisms for growth twins in face-centered cubic metals', *Acta Materialia*, 58 (2010) 2262–2270.
43. Y. Zhong, D. Ping, X. Song and F. Yin, 'Determination of grain size by XRD profile analysis and TEM counting in nano-structured Cu', *Journal of Alloys and Compounds*, 476 (2009) 113–117.
44. Z. Lee, F. Zhou, R.Z. Valiev, E.J. Lavernia and S.R. Nutt, 'Microstructure and microhardness of cryomilled bulk nanocrystalline Al–7.5% Mg alloy consolidated by

high pressure torsion', *Scripta Materialia*, 51 (2004) 209–214.

45. D. Witkin, Z. Lee, R. Rodriguez, S. Nutt and E.J. Lavernia, 'Al–Mg alloy engineered with bimodal grain size for high strength and increased ductility', *Scripta Materialia*, 49 (2003) 297–302.
46. Y.H. Zhao, T. Topping, J.F. Bingert, J.J. Thornton, A.M. Dangelewicz, Y. Li, W. Liu, Y.T. Zhu, Y.Z. Zhou and E.J. Lavernia, 'High tensile ductility and strength in bulk nanostructured nickel', *Advanced Materials*, 20 (2008) 3028–3033.
47. Y. Wang, M. Chen, F. Zhou and E. Ma, 'High tensile ductility in a nanostructured metal', *Nature*, 419 (2002) 912–915.
48. G. Dirras, J. Gubicza, S. Ramtani, Q.H. Bui and T. Szilágyi, 'Microstructure and mechanical characteristics of bulk polycrystalline Ni consolidated from blends of powders with different particle size', *Materials Science and Engineering*, A 527 (2010) 1206–1214.
49. C. Pantea, J. Gubicza, T. Ungár, G.A. Voronin and T.W. Zerda, 'Dislocation density and graphitization of diamond crystals', *Physical Review*, B 66 (2002) 094106.
50. C. Pantea, J. Gubicza, T. Ungár, G.A. Voronin, N.H. Nam and T.W. Zerda, 'High-pressure effect on dislocation density in nanosize diamond crystals', *Diamond and Related Materials*, 13 (2004) 1753–1756.
51. H. Hirai and K. Kondo, 'Shock-Compacted Si3N4 Nanocrystalline Ceramics: Mechanisms of Consolidation and of Transition from α-to β-form', *Journal of the American Ceramic Society*, 77 (1994) 487–492.
52. J. Gubicza, S. Nauyoks, L. Balogh, J. Lábár, T.W. Zerda and T. Ungár, 'Influence of sintering temperature and pressure on crystallite size and lattice defect structure in

nanocrystalline SiC', *Journal of Materials Research*, 22 (2007) 1314–1321.

53. X.J. Ning and P. Pirouz, 'A large angle convergent beam electron diffraction study of the core nature of dislocations in 3 C-SiC', *Journal of Materials Research*, 11 (1996) 884–894.
54. T. Ungár, G. Tichy, J. Gubicza and R.J. Hellmig, 'Correlation between subgrains and coherently scattering domains', *Powder Diffraction*, 20 (2005) 366–375.
55. K. Koumoto, S. Takeda, C.H. Pai, T. Sato and H. Yanagida, 'High-Resolution Electron Microscopy Observations of Stacking Faults in β-SiC', *Journal of the American Ceramic Society*, 72 (1989) 1985–1987.
56. T. Yamamoto, H. Kitaura, Y. Kodera, T. Ishii, M. Ohyanagi and Z.A. Munir, 'Consolidation of Nanostructured β-SiC by Spark Plasma Sintering', *Journal of the American Ceramic Society*, 87 (2004) 1436–1441.
57. Y.T. Zhu, X.Z. Liao, S.G. Srinivasan and E.J. Lavernia, 'Nucleation of deformation twins in nanocrystalline face-centered-cubic metals processed by severe plastic deformation', *Journal of Applied Physics*, 98 (2005) 034319.
58. L. Balogh, G. Ribárik and T. Ungár, 'Stacking faults and twin boundaries in fcc crystals determined by x-ray diffraction profile analysis', *Journal of Applied Physics*, 100 (2006) 023512.
59. T. Yamamoto, H. Kitaura, Y. Kodera, T. Ishii, M. Ohyanagi and Z.A. Munir, 'Consolidation of Nanostructured β-SiC by Spark Plasma Sintering', *Journal of the American Ceramic Society*, 87 (2004) 1436–1441.
60. T. Ungár, A. Borbély, G.R. Goren-Muginstein, S. Berger and A.R. Rosen, 'Particle-size, size distribution and dislocations in nanocrystalline tungsten-carbide', *Nanostructured Materials*, 11 (1999) 103–113.

5

Correlation between defect structure and mechanical properties of nanocrystalline materials

Abstract: The defect-related mechanical properties of nanomaterials are overviewed. The influence of small grain size on plastic deformation mechanisms, strength and ductility is discussed. The smaller the grain size in face-centred cubic (fcc) metals, the higher the activity of twinning at the expense of dislocation glide, while hexagonal close-packed (hcp) materials behave contrarily. The relationship between the dislocation structure and the yield strength is studied in detail. Above the grain size of ~20 nm, the decrease of grain size is accompanied by an increase of yield strength and a decrease of ductility. The loss of ductility can be moderated by the incorporation of coarse grains into the nanocrystalline matrix. A combination of large strength with good ductility can be achieved with nanograins containing high density of twin boundaries. Below the grain size of ~20 nm, the yield strength is found to decrease with the reduction of grain size, or remains unchanged with the experimental error. The possible explanations of this phenomenon are discussed in detail.

Key words: grain size, dislocation density, partial dislocation, twinning, yield strength, inverse Hall-Petch behaviour, ductility.

5.1 Effect of grain size on deformation mechanisms in fcc and hcp nanomaterials

In conventional coarse-grained metallic materials, the deformation is mainly controlled by the motion of full lattice dislocations in the grain interiors. Grain boundaries act as obstacles against dislocation motion, therefore the dislocations formed at Frank-Read sources inside the grains are accumulated in pile-ups, e.g. at grain boundaries. The dislocations in pile-ups apply repulsive forces to successive dislocations emitted by the sources. As the grain size decreases, the sources are located closer to the grain boundaries; therefore the emission of dislocations from the Frank-Read sources becomes more difficult. Thus, for very small grain sizes (10–20 nm) dislocation pile-ups do not form and the dislocation sources are located in the grain boundaries instead. In this case, the dislocations emitted from a boundary travel throughout the grain and are annihilated in the opposite boundary. Experiments and theoretical modelling revealed that when the grain size decreases below ~10 nm, rather partial dislocations are emitted from grain boundaries instead of perfect lattice dislocations [1–7]. Nanostructured metals produced by SPD techniques often have non-equilibrium grain boundaries that contain excess dislocations (referred to as extrinsic or non-geometrically necessary or statistically stored dislocations) in addition to the geometrically necessary ones. Some dislocations may dissociate into pairs of Shockley partials, which could move away from the grain boundary under a stress. Such grain boundaries thus act as partial dislocation sources. However, partial dislocations can also be emitted from grain boundaries by atomic reshuffling [2]. The effect of grain size on the emission of partial dislocations from grain boundaries and the related deformation twinning has been studied for fcc nanostructures on the model depicted in Figures 5.1–5.3 [8, 9]. The model assumes a grain with a square (111) slip plane, as shown in Figure 5.1. Under an external shear stress τ, a leading Shockley partial dislocation with the Burgers vector b_1 is emitted from the grain boundary XY. This leading partial consists of the three segments XE, EF and FY in Figure 5.1. τ is oriented at an angle β with line EF. The ends of this partial dislocation are pinned at the triple junctions X and Y, so that two segments of partial dislocation lines XE and FY are deposited on grain boundaries. Similarly, a trailing Shockley partial dislocation with the Burgers vector b_2 is emitted from the grain boundary XY. This partial consists of the segments XC, CD and DY. The two partial segments EF and CD are separated by a stacking fault ribbon. The two partials react to form two perfect edge dislocation segments XC and YD with the Burgers vector $b = b_1 + b_2$ at grain boundaries. The partial segments EF and CD would form a perfect screw dislocation with the Burgers vector b if they reacted. For the formation of a single leading partial, τ has to perform the work needed to increase both the area of the stacking fault behind the partial and the length of the dislocation segments XE and FY. Using the formulas for the forces acting on the leading partial dislocation segment (see Equations 3.2, 3.4 and 3.6 in Chapter 3), the critical stress, τ_P, for moving the leading partial XEFY can be derived as [9]:

$$\tau_P = \frac{1}{\cos(\beta - 30^\circ)}\left[\frac{\sqrt{6}\gamma}{a} + \frac{Ga\cdot(4-\nu)}{8\sqrt{6}\pi(1-\nu)\cdot d}\ln\frac{\sqrt{2}d}{a}\right], \qquad (5.1)$$

where G is the shear modulus, a is the lattice parameter, ν is the Poisson's ratio, γ is the stacking fault energy (SFE) and d is the grain size.

Figure 5.1 A schematic illustration of the model of a perfect screw dislocation emitted from grain boundary XY and dissociated into two partials in the slip plane (111). The stacking fault (SF) ribbon between the partials is indicated by grey colour

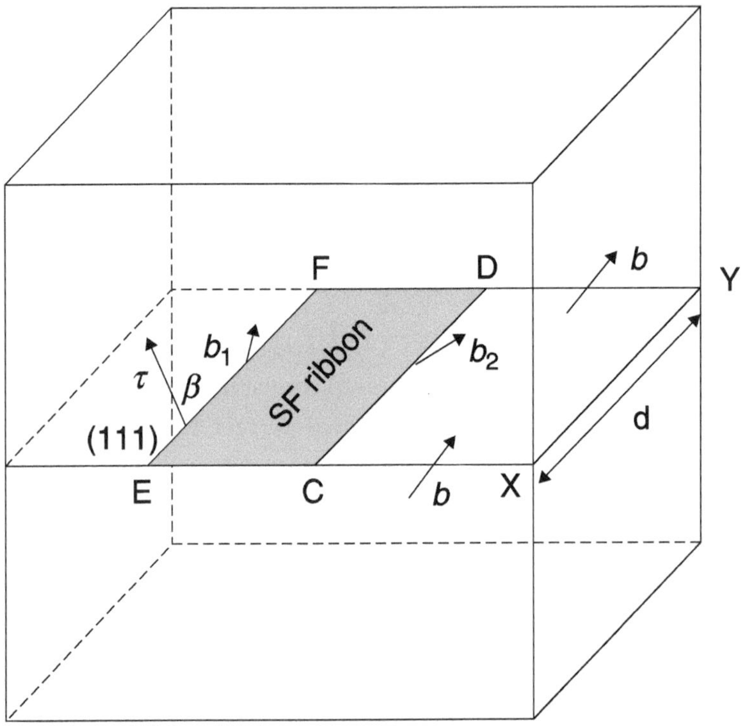

Source: After Zhu et al. (2012) [9] with permission from Elsevier.

The shear stress, τ_L, needed to move the leading and trailing partials together is equivalent to the stress for moving a full screw lattice dislocation. τ_L has to overcome the work needed to lengthen the lattice dislocation segments XC and YD, and can be obtained as [9]:

$$\tau_L = \frac{Ga}{2\sqrt{2}\pi(1-\nu)\cdot d\cos\beta}\ln\frac{\sqrt{2}d}{a}. \qquad (5.2)$$

When $\tau_P < \tau_L$, the deformation is performed by gliding of partials. For instance, in Al for $\beta = 135°$ and below the grain size of 17 nm, τ_P is smaller than τ_L [8]. Moreover, the slipping of partial dislocations on adjacent planes may yield twin nucleation. After formation of a stacking fault by slipping of a partial dislocation across the grain, a twin may nucleate via the emission of a second leading partial from the grain boundary on a plane adjacent to the stacking fault (see Figure 5.2). After slipping of the two partials on adjacent planes, the resulted stacking sequence is shown by letters A, B and C in Figure 5.2. The sequence of the (111) planes reveals that a thin twin lamella has been formed whose boundaries are also indicated in Figure 5.2. The critical twin nucleation stress, τ_{twin} can be expressed as [9]:

Figure 5.2 **A schematic illustration of the dislocation model for the nucleation of deformation twins. The stacking sequence of the (111) planes is indicated by the letters A, B and C. SF: stacking fault**

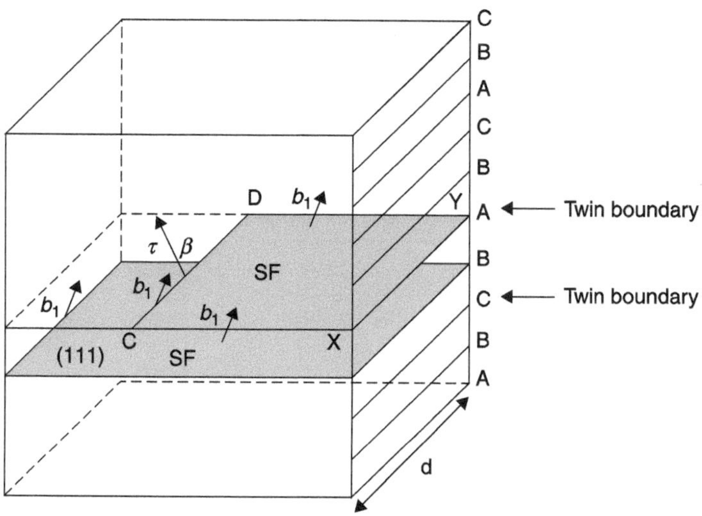

Source: After Zhu et al. (2012) [9] with permission from Elsevier.

$$\tau_{\text{twin}} = \frac{Ga \cdot (4 - \nu)}{8\sqrt{6}\pi(1 - \nu) \cdot d \cos(\beta - 30°)} \ln \frac{\sqrt{2}d}{a}. \tag{5.3}$$

Another possible scenario is that after the slipping of the first leading partial across the grain, a trailing partial is emitted from the boundary XY, which slips on the stacking fault plane and erases the stacking fault in its path, as can be seen in Figure 5.3.

Figure 5.3 A schematic illustration of the emission of a trailing partial from grain boundary XY that partially removes the stacking fault formed previously by passing a leading partial

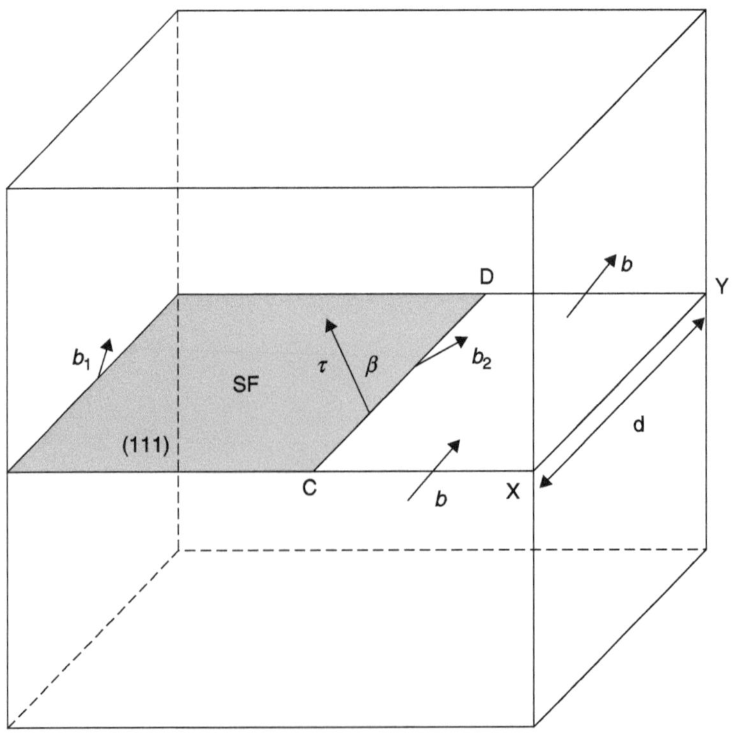

Source: After Zhu et al. (2012) [9] with permission from Elsevier.

The trailing partial requires the following stress to move [9]:

$$\tau_{\text{trail}} = \frac{1}{\cos(\beta + 30^\circ)} \left[\frac{\sqrt{6}(8+\nu)\cdot Ga}{48\pi(1-\nu)\cdot d} \ln\frac{\sqrt{2}d}{a} - \frac{\sqrt{6}\gamma}{a} \right]. \quad (5.4)$$

Once a twin is nucleated, it may grow via the emission of more twinning partials on the adjacent planes under stress τ_{twin}. However, the twin may shrink via the emission of a trailing partial with the Burgers vector b_2 on a plane adjacent to the twin boundary inside the twin (not in the parent grain). The stress needed to move a shrinking partial can be expressed as [9]:

$$\tau_{\text{shrink}} = \frac{1}{\cos(\beta + 30^\circ)} \frac{\sqrt{6}\cdot(8+\nu)\cdot Ga}{48\pi(1-\nu)\cdot d} \ln\frac{\sqrt{2}d}{a}. \quad (5.5)$$

As an example, Figure 5.4 shows the grain size dependence of the stresses τ_P, τ_L, τ_{twin}, τ_{trail} and τ_{shrink} for nanocrystalline Al. The stresses are calculated using $G = 26$ GPa, $a = 0.405$ nm, $\nu = 0.34$, $\gamma = 166$ mJ/m^2 and $\beta = 30°$. The point M in Figure 5.4 represents the critical grain size ($d_C \approx 4.7$ nm) below which a deformation twin nucleates because $\tau_{\text{twin}} < \tau_{\text{trail}}$. However, when a deformation twin nucleates by partial dislocation slipping on a plane (111), an initial stacking fault is needed on the adjacent plane. Figure 5.4 shows that at the critical grain size ~4.7 nm $\tau_L < \tau_P$, therefore the stress corresponds to the point N yields the operation of full lattice dislocation but it is not enough for the emission of partial dislocation from grain boundary that is necessary for the formation of the initial stacking fault. At the same time, for the values of τ_L at point N and d_C, the splitting distance between partials in a lattice dislocation (the stacking fault width) reaches the grain size [8]. This means that the stacking fault ribbon of a dissociated full dislocation spreads across the grain and this will serve as an initial stacking fault for the nucleation of a deformation twin. It should be emphasised that although the twin nucleation at the critical grain size d_C requires a high stress corresponding to point N, the nucleated twin can grow under a much smaller stress corresponding to point M in Figure 5.4 [9]. It is noted that the grain size dependence of the stresses τ_P, τ_L, τ_{twin}, τ_{trail} and τ_{shrink} is strongly influenced by their orientation to the possible slip systems, i.e. by β, therefore the deformation mechanism may change from grain to grain in a nanocrystalline material. Experiments also show that the operation of partial dislocations may produce deformation twins, even in nanostructured aluminium having high SFE [3–5], which in its coarse-grained state never deforms by twinning except at crack tips. For materials with lower SFE, the critical grain size and the stress required for twinning are increased and decreased, respectively. For

Figure 5.4 The shear stresses τ_P, τ_L, τ_{twin}, τ_{trail} and τ_{shrink} as a function of the grain size, *d*, for nanocrystalline Al. The angle between the stress direction and the line of slipping dislocations is β=30°

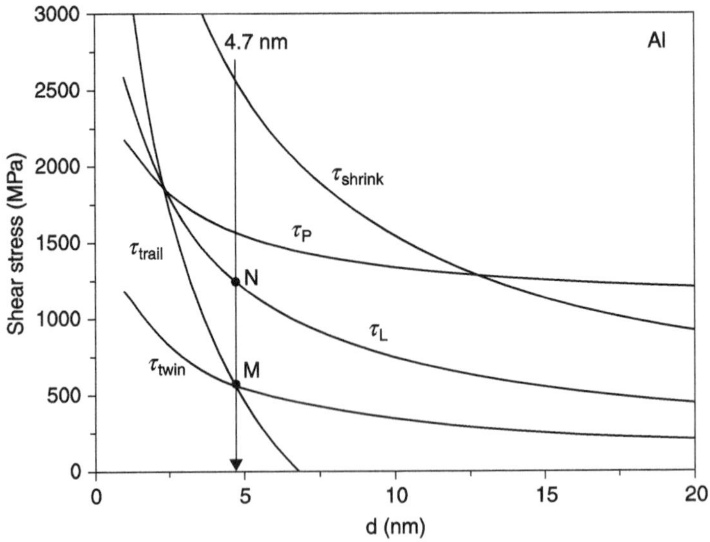

instance, while in the case of Al the critical grain size and the stress for twin nucleation are ~5 nm and ~1,000 MPa, respectively, these values for Cu are 46 nm and 370 MPa, respectively [9].

The probability of twinning in hcp structures decreases with decreasing grain size [10], which can be explained by the stress dependence of deformation mechanisms. In an hcp metal, the strain along the basal plane results from the easy glide of dislocations with <a>-type Burgers vectors. At the same time, the deformation in the direction of the crystallographic c-axis may be performed by glide of dislocations with <c+a>-type Burgers vectors or by twinning (both processes operate mainly on pyramidal planes). The most frequently observed twinning planes and directions are $\{10.1\}\langle 10.\bar{2}\rangle$ and $\{11.2\}\langle 11.\bar{3}\rangle$ (compressive twins), and $\{10.2\}\langle 10.1\rangle$ and $\{11.1\}\langle \bar{1}\bar{1}.6\rangle$ (tensile twins). The generation and propagation of <c+a> dislocations are difficult due to their large Burgers vectors. However, an elevated deformation temperature and/or high stresses increase the activation of pyramidal <c+a> dislocations at the expense of twin formation. In deformed hcp metals with smaller grain size, the stresses are higher mainly in the vicinity of grain boundaries [11], which facilitate the operation of <c+a> dislocations in addition to <a> dislocations, thereby reducing the occurrence of twinning. It is noted that deformation twins are observed in nanocrystalline Mg-10 at.% Ti alloy with an average grain size of 33 nm processed by mechanical attrition. The formation of deformation twins is attributed to the alloying effect, which may change the energy path for twinning [12].

As the grain boundary fraction in nanomaterials increases, the relative importance of deformation mechanisms occurring at grain boundaries increases. Such deformation mechanism is grain boundary sliding controlled by grain boundary diffusion. Grain boundary diffusion is much faster than crystal lattice diffusion inside grains, since the activation energy of the former process is about half of that for the latter mechanism. However, in coarse-grained materials (*d* > 10 μm) the contribution of grain-boundary diffusion to plastic deformation (Coble creep) is marginal due to the small volume fraction of grain boundaries. For these materials, lattice diffusion inside grains is fast enough for controlling plastic deformation (Nabarro-Herring creep) only at temperatures above $0.5 \times T_m$, where T_m is the absolute melting temperature. At the same time, in nanomaterials grain-boundary sliding by Coble creep may operate even at room temperature (RT) as a very high fraction of atoms is located in the vicinity of grain boundaries. Figure 5.5 shows the Ashby-Verrall model of grain-boundary sliding [13, 14]. The grains are sliding at their boundaries due to the external tensile stress. The sliding of the grains without any change of their shape would yield voids between them, therefore the integrity of the material is sustained by a complementary diffusion in the vicinity of the grain boundaries, as shown in the bottom part of Figure 5.5. The combination of sliding and diffusion yields a strain of 0.55 in this model. For metallic materials with high melting points, grain-boundary sliding may contribute to plasticity at RT even if the grain size is only in the ultrafine-grained (UFG) regime, i.e. between 100 nm and 1 μm. For instance, atomic-force microscopy (AFM) study shows rumpling on the surface around a Vickers pattern in pure Al processed by equal-channel angular pressing (ECAP) and having a grain size of about 1 μm [15, 16]. The quantitative evaluation of the AFM measurements reveals that adjacent grains may slide up to ~30 nm over each other and the contribution of grain-boundary sliding to the total strain is about ~70% during indentation of the ECAP-processed Al sample. Other experimental investigations [17–19] and three-dimensional molecular

Figure 5.5 Schematic illustration of the Ashby-Verrall model for grain-boundary sliding. The arrows in the grey grains indicate the sliding directions along the grain boundaries. At the bottom of the figure, the arrows show the directions of diffusion in the vicinity of grain boundaries, which yields the change of the grain shape as a complementary process in addition to grain-boundary sliding

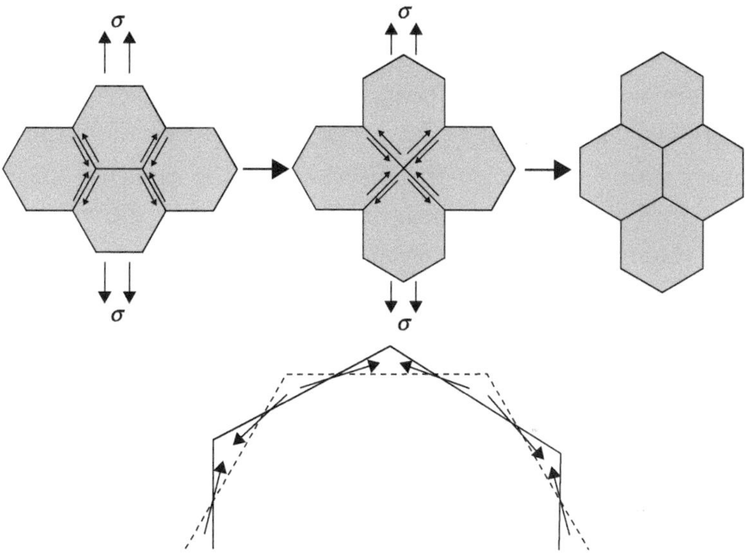

dynamics computer simulations [20, 21] also suggest the occurrence of grain-boundary sliding as a viable deformation process in nanocrystalline solids at low temperatures. In addition to grain-boundary sliding, other boundary-related deformation mechanisms, such as grain rotation, are also observed in nanomaterials [22]. During this process, nanosized grains rotate in a manner that brings their orientations closer together. The disappearance of the boundary between these grains provides a path for more extended dislocation motion [23].

5.2 Breakdown of Hall-Petch behaviour in nanomaterials

Considering the grain boundary strengthening in polycrystalline coarse-grained metallic materials, the yield strength is usually correlated to the grain size by applying the Hall-Petch equation [24–26]. In a grain of a polycrystal, the plastic deformation sets in when the resolved shear stress in one or more slip systems reaches the critical value necessary for the initiation of dislocation glide. In a polycrystalline material, the grains have different crystallographic orientations relative to the direction of the external load, therefore the resolved shear stresses acting on the various slip systems also change from grain to grain. In the grains with favourable orientations (i.e. having slip systems with large Schmid-factors), the resolved shear stresses are high and the plastic deformation starts by the activation of one or more slip systems. In unfavourably oriented grains, the shear stresses are not enough to initiate dislocation glide, therefore these grains deform only elastically. In the plastically deformed grains, dislocations are formed at the Frank-Read sources and these dislocations are accumulated in pile-ups at the grain boundaries as depicted for grain A in Figure 5.6. The elastic stresses of dislocations in the pile-ups are added to the shear stresses caused by the external load, thereby assisting the initiation of plastic deformation in the adjacent, elastically deformed grain having unfavourable orientation relative to the applied force (grain B in Figure 5.6). The macroscopic plastic deformation sets in when all grains start to deform plastically. As the repulsive forces between dislocations in pile-ups are inversely proportional to dislocation spacing, therefore the smaller the grain size, the lower the number of dislocations in the pile-ups. As a consequence, when the grain size is reduced, the contribution of the stress field of dislocations to the initiation of plasticity in the adjacent grain is smaller, therefore larger applied stress is necessary for yielding the material, i.e. the yield strength will be higher. Thus, there is an inverse relationship between grain size and yield strength, as demonstrated by the Hall-Petch equation [24–26]:

$$\sigma_Y = \sigma_0 + k_Y \cdot d^{-1/2}, \tag{5.6}$$

Figure 5.6 Two grains A and B with the size *d* in a polycrystalline material loaded by a tensile stress σ. The dashed lines represent the slip planes. In grain A, the plastic deformation has already been initiated by the external stress and a Frank-Read source (FR) emits dislocations that are gliding due to the shear stress τ. The stress field of the dislocations accumulated at the boundary in grain A assists the activation of Frank-Read sources in the unfavourable oriented grain B

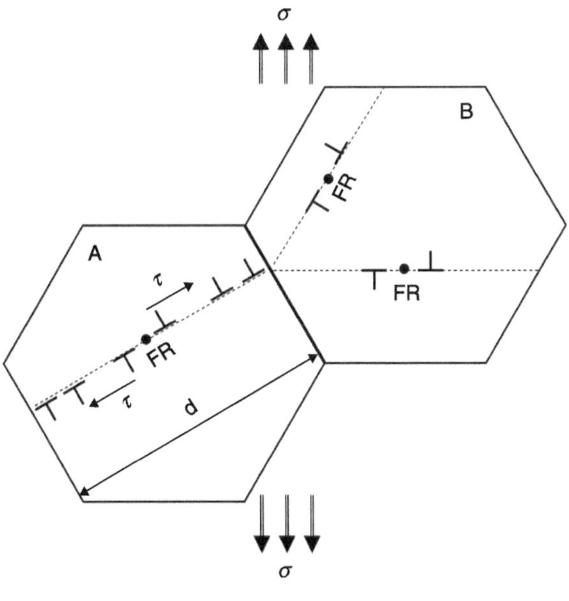

where σ_Y is the yield strength, σ_0 is the friction stress, k_Y is the strengthening coefficient (a constant unique to each material) and d is the average grain diameter. For instance, the values of σ_0 and k_Y are ~25 MPa and ~5,000 MPa·nm$^{1/2}$ for Cu [26] as illustrated in Figure 5.7.

When the grain size is reduced to the range 20–100 nm, the yield strength further increases with decreasing grain size but the slope k_Y continuously decreases (see Figure 5.7) [23]. This deviation from the conventional Hall-Petch behaviour can be attributed to the grain size dependence of the stress (τ) required to operate a Frank-Read source [32, 33]. If an edge dislocation segment pinned at its ends is considered as a

Figure 5.7 The Hall-Petch plot of the yield strength (σ_Y) versus the grain size (d) for Cu

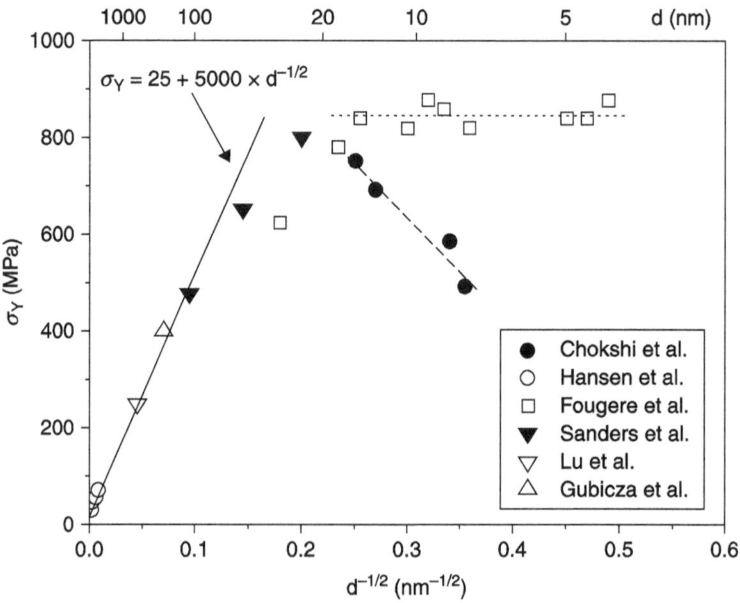

Sources: Chokshi et al. (1989) [27], Hansen and Ralph (1982) [26], Fougere et al. (1992) [28], Sanders et al. (1997) [29], Lu et al. (2005) [30] and Gubicza et al. (2009) [31].

Frank-Read source inside a grain and $\nu = 0.33$ is chosen for the Poisson's ratio, the stress τ versus the length of the segment L can be given as:

$$\tau = \frac{0.12Gb}{L}\left[\ln\left(\frac{L}{b}\right) - 1.653\right], \tag{5.7}$$

where b is the magnitude of the Burgers vector and G is the shear modulus [33]. τ as a function of $L^{-1/2}$ for Cu is plotted in Figure 5.8. When the length of the dislocation segment is large, the stress needed for the operation of a source increases more or less linearly with $L^{-1/2}$. However, when the length of a Frank-Read source is reduced below 10–15 nm, the slope of τ versus $L^{-1/2}$ plot decreases. Moreover, if L is smaller

Figure 5.8 The stress (τ) required for the operation of a Frank-Read source versus the length (L) of the edge dislocation segment between the pinning points of the source in Cu

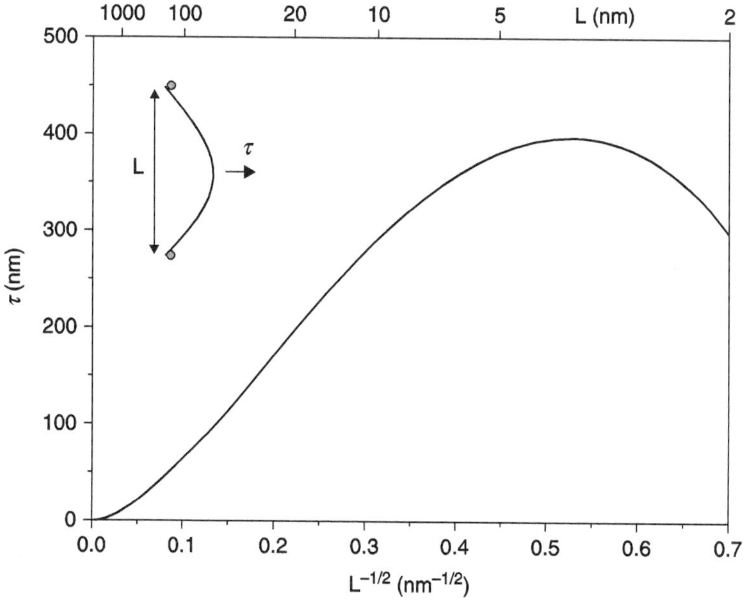

than 4 nm in Cu, the stress required for the operation of a Frank-Read source decreases with the reduction of its length. In materials with grain sizes between 20 and 100 nm, a considerable fraction of Frank-Read sources may have smaller lengths than 10–15 nm, which can explain the deviation from the conventional Hall-Petch behaviour. Additionally, the strong reduction of the numbers of dislocations in pile-ups with decreasing grain size yields a decrease of the exponent of d in Equation 5.6 from $-1/2$ to -1, which may cause the observed reduction of k_Y in the Hall-Petch plot [16]. The value of -1 for the Hall-Petch exponent is also predicted by the model in which two dislocations locate within one grain with the distance of $d/3$ between them, where d is the grain diameter [34].

As the grain size decreases to 10–20 nm, the number of dislocations in pile-ups is eventually reduced to one and thus the pile-up mechanism breaks down [23]. As a consequence, the assisting effect of the stress field of dislocation pile-ups on the initiation of plastic deformation is lost and the Hall-Petch equation is not valid. In this grain size regime, if dislocations contribute to plasticity, they are emitted from grain boundaries, then travel through the grains with a small chance of interaction with other dislocations and finally annihilate at the opposing boundaries. Below 10–20 nm, the experimental results do not show a clear trend in the grain size dependence of the yield strength. In some cases, the yield strength is found to decrease with the reduction of grain size while in other cases only a plateau is observed (see Figure 5.7). When the yield strength decreases with decreasing grain size, the Hall-Petch plot remains more or less linear even below the grain size of 10–20 nm but with a negative slope, therefore this behaviour is termed as negative or inverse Hall-Petch behaviour. This effect is usually attributed to the dominance of deformation mechanisms occurring in the grain boundaries as the reduced grain size is accompanied by the increase of grain boundary fraction. Such a mechanism is grain-boundary sliding that may occur by grain-boundary diffusion controlled Coble creep [35]. In this case, the stress (σ) required for yielding at a strain rate of $\dot{\varepsilon}$ can be given by the following formula [23]:

$$\sigma = \frac{\pi k T \dot{\varepsilon}}{150 \Omega \delta D_{GB}} d^3, \qquad (5.8)$$

where k is Boltzmann's constant, T is the absolute temperature, Ω is the atomic volume, δ is the grain boundary width and D_{GB} is the grain-boundary diffusion coefficient. Assuming $\Omega = 1.3 \times 10^{-29}$ m^3, $\delta = 1$ nm, $D_{GB} = 4.7 \times 10^{-20}$ m^{2}s^{-1} at RT and $\dot{\varepsilon} = 10^{-4}$ for Cu with $d = 20$ nm, the stress required for Coble creep is 112 MPa as obtained from Equation 5.8, which is much less than the calculated stress needed for dislocation glide (~800–900 MPa). The easy occurrence of Coble creep suggests its significant contribution to plasticity at RT for grain sizes less than 20 nm. Moreover, Equation 5.8 predicts that when the grain-boundary sliding becomes the main deformation mechanism, the yield strength will decrease with decreasing grain size, a trend that corresponds to a considerable fraction of experimental results. However, Coble creep cannot be the sole deformation mechanism in the inverse Hall-Petch behaviour as the experimental yield strength values do not fall so steeply with decreasing grain size as the trend, $\sigma \sim d^3$, predicted by Equation 5.8 [36]. For instance, for inert-gas condensed Cu the inverse Hall-Petch effect was described as:

$$\sigma_Y = A - k' \cdot d^{-1/2}, \qquad (5.9)$$

where $A \approx 1{,}290$ MPa and $k' \approx 2{,}170$ MPa·nm$^{1/2}$ [27]. Most probably, intragranular dislocation motion as a complementary deformation mechanism also occurs in order to sustain the integrity of the material as the sliding of grains without changing their shape would yield voids between them. It should also be noted that the voids, micro- or nano-cracks and poorly bonded interfaces in nanomaterials processed by powder consolidation (e.g. inert-gas condensation) may also yield a similar strength degradation as the inverse Hall-Petch behaviour [23]. Although the experimental results obtained on different nanomaterials do not prove convincingly the general existence of the inverse Hall-Petch effect, molecular dynamic simulations predict this behaviour for very small grain sizes [37]. It is noted, however, that the results of molecular dynamic simulations are valid only for very high strain rates (10^8–10^9 s^{-1}). Moreover, the change of the deformation mechanism inside the grains when their sizes decrease below ~20 nm may also contribute to the inverse Hall-Petch behaviour. This is illustrated in Figure 5.9, where the shear stresses required for the emission of full lattice (τ_L), leading partial (τ_P), trailing partial (τ_{trail}), twinning partial (τ_{twin}) and shrinking partial (τ_{shrink}) dislocations from grain boundaries are plotted as a function of grain size in Cu. These stresses are calculated according to Equations 5.1–5.5 assuming G=47 GPa, a=0.361 nm, ν=0.34, γ=62 mJ/m^2 and β=30°. It seems that below ~30 nm twinning becomes a considerable deformation mechanism. For twin nucleation, first an initial stacking fault should be formed at the stress corresponding to point *N*. Then, twinning occurs at the stress corresponding to point *M*, which is much lower than that at point *N*. When the grain size falls below ~20 nm (point *L* in Figure 5.9), partial dislocations are emitted from the grain boundaries instead of full dislocations since the former mechanism requires smaller stress. In nanocrystalline materials, the grains have a size distribution. With decreasing the average grain size, more and more grains in the distribution become so small that their interiors deform rather by twinning, which requires smaller stress than for dislocation glide in the large

Figure 5.9 The shear stresses τ_P, τ_L, τ_{twin}, τ_{trail} and τ_{shrink} calculated from Equations 5.1–5.5 as a function of the grain size, *d*, for nanocrystalline Cu. The angle between the stress direction and the line of slipping dislocations is $\beta = 30°$

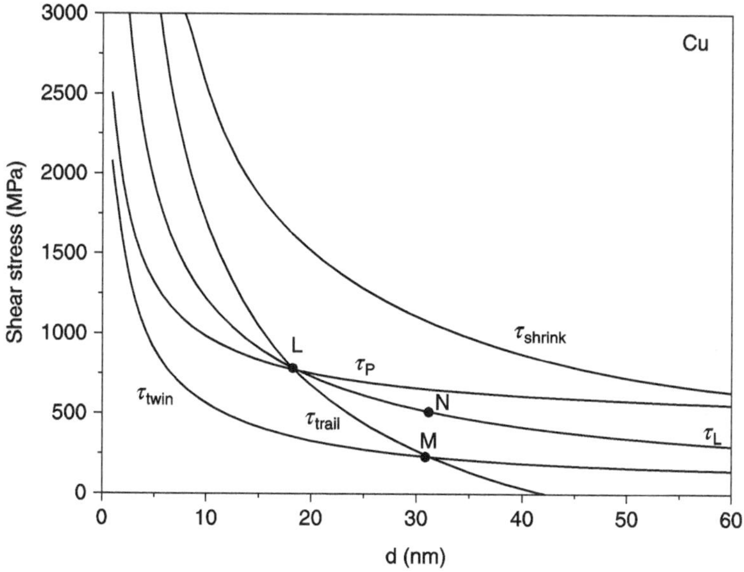

nanograins. Therefore, the macroscopic yield strength decreases with decreasing the average grain size in accordance with the inverse Hall-Petch behaviour.

5.3 Correlation between dislocation structure and yield strength of ultrafine-grained fcc metals and alloys processed by severe plastic deformation

For UFG fcc metals processed by ECAP, the dislocation density and the yield strength show similar developments as a function of strain and both quantities saturate at approximately the same strain [16, 38]. This suggests that it is worthwhile studying the relationship between these quantities. The dislocation density determined by x-ray line profile analysis and the corresponding yield strength measured by mechanical tests for different fcc metals and alloys are listed in Table 5.1. The relationship between the dislocation density (ρ) and yield strength (σ_Y) for plastically deformed metals is generally characterised by the Taylor equation:

$$\sigma_Y = \sigma_0 + \alpha M^T G b \rho^{1/2}, \quad (5.10)$$

where σ_0 is the friction stress, α is a constant depending on the arrangement of dislocations and M^T is the Taylor factor. Since the value of αM^T is of the order of 1, it is possible to check the validity of Equation 5.10 by plotting the value of $(\sigma_Y - \sigma_0)$ against the product of $Gb\rho^{1/2}$ for different fcc metals and solid solution alloys processed by SPD, as shown in Figure 5.10. The good correlation between these two quantities in Figure 5.10 demonstrates that in UFG fcc metals and solid solutions processed by SPD the yield strength is essentially determined by the interactions between dislocations. This observation can be explained by the fact that in materials processed by SPD at low homologous temperatures, the grain refinement occurs mainly by the arrangement of dislocations into subgrain-boundaries and/or dislocation walls. Consequently, a large fraction of boundaries consists of dislocations and the effect of subgrain/grain boundaries on dislocation motion can be regarded as the interaction between dislocations [39]. Concerning the solid solution strengthening, the solute atoms may increase the strength in two ways: (i) directly by pinning the dislocations and impeding their motion in a solute–dislocation interaction and (ii) indirectly by hindering the annihilation of dislocations during deformation, leading to an increase in the dislocation

Table 5.1 The dislocation density (ρ) determined by x-ray line profile analysis and the yield strength (σ_Y) at RT for fcc metals and alloys processed by SPD. The calculated values of α in the Taylor equation (Equation 5.10) for pure metals and solid solutions are also presented. G: shear modulus, b: Burgers vector, σ_0: friction stress, HPT: high-pressure torsion

Material	Processing method	ρ [10^{14} m^{-2}]	σ_Y [MPa]	σ_0 [MPa]	G [GPa]	b [nm]	α	Reference
			Pure metals					
Al	1 ECAP at RT	1.9	95	20	26	0.286	0.24	[40]
Al	2 ECAP at RT	2.1	110	20	26	0.286	0.27	[40]
Al	4 ECAP at RT	1.9	120	20	26	0.286	0.32	[40]
Al	8 ECAP at RT	1.8	120	20	26	0.286	0.33	[40]
Cu	1 ECAP at RT	15	294	35	47	0.256	0.18	[31]
Cu	3 ECAP at RT	21	386	35	47	0.256	0.21	[31]
Cu	5 ECAP at RT	21	394	35	47	0.256	0.21	[31]
Cu	10 ECAP at RT	20	412	35	47	0.256	0.23	[31]
Cu	15 ECAP at RT	15	404	35	47	0.256	0.26	[31]
Cu	25 ECAP at RT	15	402	35	47	0.256	0.26	[31]
Cu	25 HPT at RT	37	563	35	47	0.256	0.24	[41]
Au	4 ECAP at RT	17	245	27	27	0.288	0.22	[42]
Ni	6 ECAP	18	730	60	82	0.249	0.25	[42]

(Continued)

Table 5.1 The dislocation density (ρ) determined by x-ray line profile analysis and the yield strength (σ_Y) at RT for fcc metals and alloys processed by SPD (*Continued*)

Material	Processing method	ρ [10^{14} m^{-2}]	σ_Y [MPa]	σ_0 [MPa]	G [GPa]	b [nm]	α	Reference
Pure metals								
Ni	8 ECAP + 5 HPT at RT	25	1103	60	82	0.249	0.33	[40]
Ag	8 ECAP at RT	46	330	29	30	0.289	0.17	[42]
Solid solution alloys								
Al-1Mg	8 ECAP at RT	3.9	220	25	26	0.286	0.43	[40]
Al-3Mg	1 ECAP at RT	12	250	50	26	0.286	0.25	[40]
Al-3Mg	8 ECAP at RT	23	400	50	26	0.286	0.32	[40]
Cu-10%Zn	5 HPT at RT	34	773	35	46	0.256	0.35	[43]
Cu-30%Zn	5 HPT at RT	81	987	44	41	0.256	0.33	[43]
Dispersion-strengthened alloys								
Al 6082	1 ECAP at RT	4.0	193	25	26	0.286	–	[40]
Al 6082	4 ECAP at RT	5.0	246	25	26	0.286	–	[40]
Al 6082	8 ECAP at RT	5.4	265	25	26	0.286	–	[40]
Al-4.8Zn-1.2Mg-0.14Zr	8 ECAP at 200°C	3.2	290	20	26	0.286	–	[44]
Al-5.7Zn-1.9Mg-0.35Cu	8 ECAP at 200°C	3.4	380	20	26	0.286	–	[44]
Al-5.9Mg-0.3Sc-0.18Zr	15 HPT at RT	24	750	20	26	0.286	–	[44]
Cu-0.18%Zr	8 ECAP at RT	39	416	35	47	0.256	–	[44]

Figure 5.10 The yield strength at RT reduced by the friction stress ($\sigma_Y - \sigma_0$) vs. the product of $Gb\rho^{1/2}$ for different fcc metals and alloys processed by SPD. The errors on the individual datum points are indicated by solid horizontal and vertical lines at the symbol diamond. *G*: shear modulus, *b*: Burgers vector, ρ: dislocation density

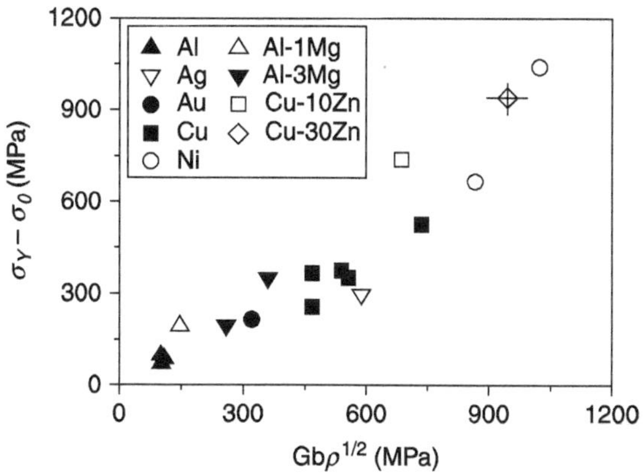

density in a dislocation–dislocation interaction. The direct and indirect effects of solute atoms are taken into account in the friction stress, σ_0, and the dislocation term in the Taylor equation, respectively. For severely deformed metals where the dislocation density is relatively high, the latter term is dominant so that the hardening effect of the solute atoms is manifested basically in the higher dislocation density.

The scattering of the datum points in Figure 5.10 is attributed to the difference between the values of α for the different metals. Therefore, the values of α in Table 5.1 are calculated from Equation 5.10 by using the experimental values of σ_Y and ρ. In this calculation, the Taylor factor is estimated by the value characteristic for a random crystallographic orientation (3.06) as weak textures in fcc metals after SPD, if any, yield only a deviation less than 3% from this value of M^T [42]. Most of the values of α are between 0.17 and 0.35, and this range is similar to that observed for coarse-grained materials. In practice, the value of α depends upon the arrangement of dislocations in the material [45, 46]. For example, calculations have shown that the value of α increases from ~0.15 to ~0.37 when dislocation clustering increases and the dislocation structure evolves from a uniform random distribution through thick cell walls to sharp boundaries [45]. The dissociation of lattice dislocations into partials in fcc metals has a strong effect on the arrangement of dislocations. In the formation of dense dislocation structures, as in walls or subgrain boundaries, an important role is played by cross-slip and climb. Thus, the higher the degree of dissociation, the more difficult the occurrence of cross-slip and climb. The degree of dislocation dissociation is characterised by the equilibrium splitting distance (d_p) between the partial dislocations, which can be determined from the shear modulus, the Burgers vector and the stacking fault energy using Equation 3.3 given in Chapter 3. The values of d_p are listed in Table 3.1 in Chapter 3 for edge and screw dislocations in pure fcc metals. The values of α obtained for pure fcc metals processed by ECAP at RT until the saturation of the dislocation density are plotted in Figure 5.11 as a function of the arithmetic average of splitting distances for edge and screw dislocations in Burgers vector unit, d_p/b. The higher the value of d_p/b, the higher the degree of dislocation dissociation, which impedes the clustering of dislocations inside the grains resulting in a relatively low value of α. It should be noted that, besides the interaction between dislocations, the relatively high twin boundary frequency in Ag (see Chapter 3) is expected to make an additional contribution to the strength. However, if this hardening effect is taken into account, the Taylor-type contribution to the total yield strength will be reduced and this will lead to an even smaller value of α for Ag, providing additional confirmation that the higher splitting distance is accompanied by a smaller value of α. In the cases of Cu-10% Zn and Cu-30% Zn, the values of α are most

Figure 5.11 The value of α in the Taylor equation as a function of the equilibrium splitting distance between partials in Burgers vector unit (d_p/b) for pure fcc metals processed by ECAP at RT until saturation of the dislocation density

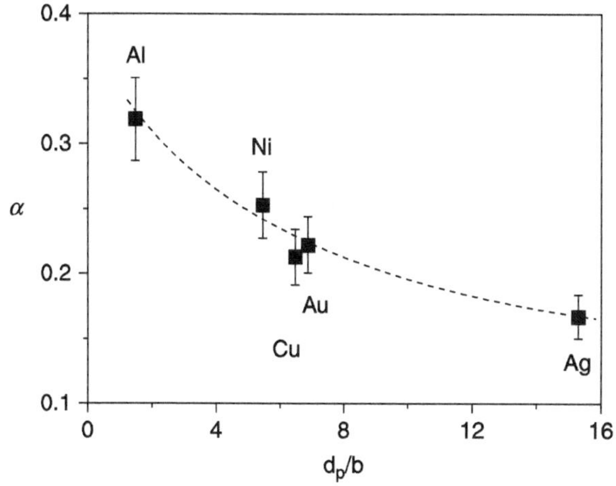

Source: Reprinted from Gubicza et al. (2009) [31] with permission from Elsevier.

probably also overestimated in Table 5.1 due to the large twin boundary hardening.

It is noted that α varies with increasing strain in SPD-processing of pure fcc metals. Figure 5.12 plots the value of α as a function of the number of ECAP passes for pure Al and Cu. The increase of the value of α with increasing strain is explained by the thinning of the grain/subgrain boundaries and the evolution to a more equilibrated structure as discussed in Chapter 2. It is worthwhile to note that in the case of Cu, α saturates only after 15 ECAP passes, although the maximum dislocation density has already been achieved after 5 passes, indicating that the evolution of the dislocation structure continues even after the saturation of its density.

The yield strength values for some dispersion-strengthened UFG alloys are also presented in Table 5.1 but for these

Figure 5.12 The value of α in the Taylor equation as a function of the number of ECAP passes for pure Al and Cu

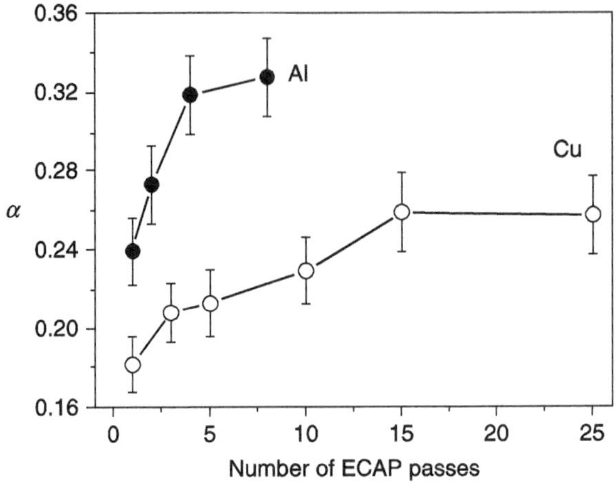

materials the value of α is not determined due to the strength contribution of the secondary phase dispersoids (precipitates). Considering non-shearable dispersoids in UFG alloys and assuming the additivity of the different strengthening contributions, the yield strength may be expressed by the following relationship [47]:

$$\sigma_y = \sigma_0 + \alpha M G b \rho^{1/2} + 0.85M \frac{Gb\,ln(x/b)}{2\pi(L-x)}, \tag{5.11}$$

where x is the average size of dispersoids and L is the average distance between them. For instance, in supersaturated Al-4.8Zn-1.2Mg-0.14Zr and Al-5.7Zn-1.9Mg-0.35Cu (wt.%) alloys processed by eight passes of ECAP at 200°C stable incoherent $MgZn_2$ precipitates (η-phase particles) form [44]. The different contributions to yield strength are calculated from Equation 5.11 using the values of L and x obtained from TEM micrographs and the dislocation density determined by x-ray line profile analysis. The average size of the precipitates in the Al-4.8Zn-1.2Mg-0.14Zr and Al-5.7Zn-1.9Mg-0.35Cu alloys are 30 and 20 nm, respectively [44]. The average distance, *L*, between the precipitates is approximately 120 nm in the Al-4.8Zn-1.2Mg-0.14Zr alloy and 80 nm in the Al-5.7Zn-1.9Mg-0.35Cu alloy. The value of σ_0 was taken as 20 MPa measured for pure Al [48] because the lattice parameter of the matrix obtained from the x-ray diffractograms agrees with the value for pure Al within the experimental error. The strength contributions originating from the dislocation density given by the second term in Equation 5.11 are 133 and 137 MPa for Al-4.8Zn-1.2Mg-0.14Zr and Al-5.7Zn-1.9Mg-0.35Cu, respectively, as the dislocation densities in the two matrices are very close to each other. The third term in Equation 5.11 related to the dislocation–precipitation interaction gives 156 and 214 MPa for Al-4.8Zn-1.2Mg-0.14Zr and Al-5.7Zn-1.9Mg-0.35Cu, respectively, where the difference between the two values is due to the different size and dispersion of precipitates [44]. Thus, the smaller particles with a more dense distribution lead to a higher strength for the Al-5.7Zn-1.9Mg-0.35Cu alloy. Finally, the sum of the three components leads to estimated total strengths of 309 and 371 MPa, which are in good agreement with the values of 290 and 350 MPa determined by mechanical testing for the Al-4.8Zn-1.2Mg-0.14Zr alloy and the Al-5.7Zn-1.9Mg-0.35Cu alloy, respectively (see Table 5.1).

5.4 Defect structure and ductility of nanomaterials

Besides the yield strength, the ductility of UFG and nanomaterials is also an important parameter describing the mechanical behaviour. The ductility is usually characterised by the maximum elongation or the maximum true strain

(ε_{max}) achieved until inhomogeneous deformation (necking) sets in during tension. After necking, the deformation is localised mainly to the neck, and due to the decrease of the cross-section at the neck, the force required for further straining decreases, i.e. the deformation becomes unstable (it does not stop immediately when the force is reduced). As a consequence of inhomogeneous deformation after necking, the true stress (σ) and true strain (ε) vary along the longitudinal axis of the specimen, therefore the following usual formulas for the calculation of σ and ε can only be applied before necking:

$$\varepsilon = \ln\frac{l}{l_0}, \tag{5.12a}$$

$$Al = A_0 l_0, \tag{5.12b}$$

$$\sigma = \frac{F}{A}, \tag{5.12c}$$

where l and l_0 are the actual and initial lengths of the sample, respectively, F is the applied force, A and A_0 are the actual and initial cross-sections of the sample, respectively. Equation 5.12b reflects the unchanged volume during plastic deformation. The deformation is stable only in the homogeneous regime, i.e. when the first derivative of the force over the sample length, $\frac{dF}{dl}$, is larger than zero. Using Equations 5.12a–5.12c, this criterion can be expressed by the true stress and true strain as:

$$\frac{d\sigma}{d\varepsilon} \geq \sigma. \tag{5.13}$$

This formula is referred to as Considère-criterion.

In the case of UFG and nanomaterials, the flow stress at RT is large even in the beginning of tension due to the high initial dislocation density and/or the small grain size, and it tends to saturate at small strain values. As a result, for UFG and nanomaterials the true stress, σ becomes equal to the work hardening rate, $\frac{d\sigma}{d\varepsilon}$ at smaller strains than for coarse-grained counterparts, i.e. the ductility decreases with decreasing grain size as illustrated in the schematic picture of Figure 5.13. As an example, Figure 5.14 shows the grain size dependence of ductility at RT for interstitial-free steel processed by accumulative-roll bonding (ARB) [49]. The grain size varies between 200 nm and 20 μm, and the ductility is described by the uniform and total elongations. The former and the latter quantities are determined when $\frac{dF}{dl}$ becomes zero and after the failure of the sample, respectively. Figure 5.14 reveals that a large loss of ductility occurs when the grain size decreases below 1 μm. Similar effect is also observed for Al-alloys processed by SPD [49]. In addition to the small value of work

Figure 5.13 **Schematic illustration of decreasing ductility (ε_{max}) with decreasing grain size according to Considère-criterion. UFG: ultrafine-grained material, CG: coarse-grained material**

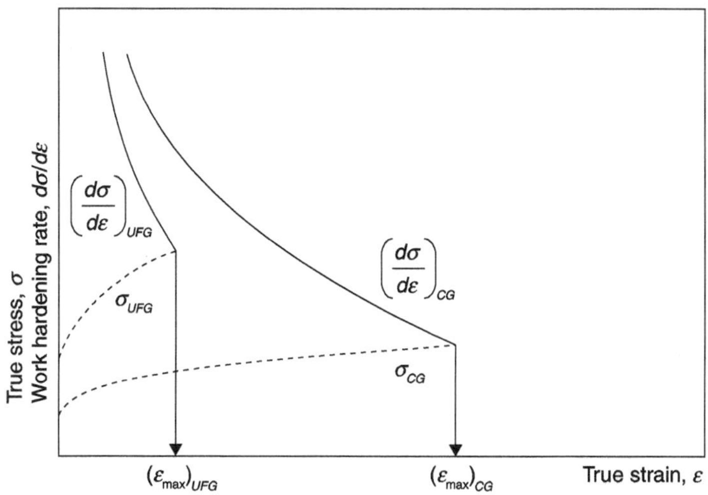

Figure 5.14 Grain size dependence of uniform and total elongations at RT for interstitial-free steel with various mean grain sizes

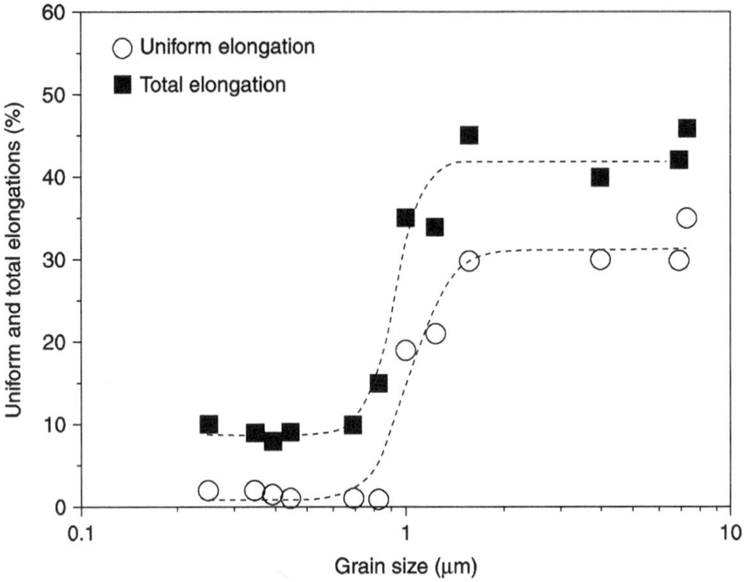

Source: The data are taken from Tsuji et al. (2002) [49].

hardening rate, the considerable porosity and amount of oxide phase at the interfaces in UFG and nanomaterials consolidated from nanopowders also reduce the ductility. The relative fractions of pores and oxide particles usually increase with decreasing grain size as a result of the larger specific surface area for a smaller initial particle size. The porosity and the oxide phase may yield weaker particle bonding, resulting in an easier cracking under external loads. Generally, in UFG and nanomaterials the total elongation is typically less than a few per cent and the regime of uniform deformation is even smaller, which limits their practical utility.

For materials whose plastic deformation is sensitive to strain-rate ($\dot{\varepsilon}$), the strain regime of homogeneous deformation is determined by Hart's stability criterion:

$$\frac{d\sigma}{d\varepsilon} \geq \sigma(1-m), \tag{5.14}$$

where m is the strain rate sensitivity parameter defined as:

$$m = \left(\frac{\partial \ln \sigma}{\partial \ln \dot{\varepsilon}}\right)_T . \tag{5.15}$$

The value of m is usually between 0 and 1, but it depends on the microstructure, the temperature and the strain rate of deformation. The higher strain rate sensitivity helps to sustain the homogeneous deformation since the material hardens at the neck when the strain rate increases in the beginning of necking, which yields larger ductility. If the value of m is larger than 0.3, very large strains (higher than 100%) can be achieved during tension and this behaviour is termed superplasticity. Superplastic deformation in coarse-grained metals and alloys is usually observed at low strain rate (10^{-5}–$10^{-2}\,s^{-1}$), high homologous temperature (>0.5 T_m, where T_m is the melting point) and for relatively small grain size (<10 μm) since its mechanism is the grain-boundary sliding. As a consequence, the decrease in grain size usually improves the superplastic behaviour. This effect is illustrated on a commercial spray-cast Al-7034 alloy having a chemical composition of 11.5% Zn-2.5%Mg-0.9% Cu-0.2% Zr (in wt.%) with the balance of Al [50, 51]. The initial grain size of 2.1 μm is reduced to 300 nm after six ECAP passes at 200°C. An as-received specimen and a sample processed by six passes were subjected to tensile testing at 350°C at strain rates between 10^{-3} and 10^{-1} s^{-1}. The elongations in the as-received condition were in the range of 150–300%. The ECAP-processed specimen showed higher elongation for any strain rate and the maximum elongation was higher than 1,000% at strain rates in the vicinity of ~10^{-2} s^{-1} [52]. The very high elongations for the ECAP-processed sample are due to the UFG microstructure, which was retained even at high testing temperatures because of the presence of finely dispersed $MgZn_2$ and Al_3Zr precipitates, which impeded grain boundary motion. In this case, ECAP-processing results in an increase of the strain-rate sensitivity from ~0.2 to 0.3–0.5 [51].

At low homologous temperature, the strain rate sensitivity parameter is usually less than 0.05, because the superplastic deformation mechanisms (e.g. grain-boundary sliding) are controlled by thermally activated diffusion. Figures 5.15a and 5.15b show that with decreasing grain size in fcc and body-centred cubic (bcc) metals, the value of m increases and decreases, respectively, which can be explained by the different nature of activation volumes of deformation [35]. The activation volume, $V*$ for a dislocation controlled deformation is defined as the product of the Burgers vector, b and the activation area for dislocation motion, A^*. The latter quantity is the area swept by a dislocation overcoming obstacles in its course. The activation volume can be expressed from the strain-rate sensitivity as:

$$V^* = \frac{3^{1/2} kT}{m\sigma}, \tag{5.16}$$

Figure 5.15 **The strain-rate sensitivity, *m*, measured at RT as a function of grain size for (a) fcc Ni [60–62] and (b) bcc Fe processed by powder metallurgy methods**

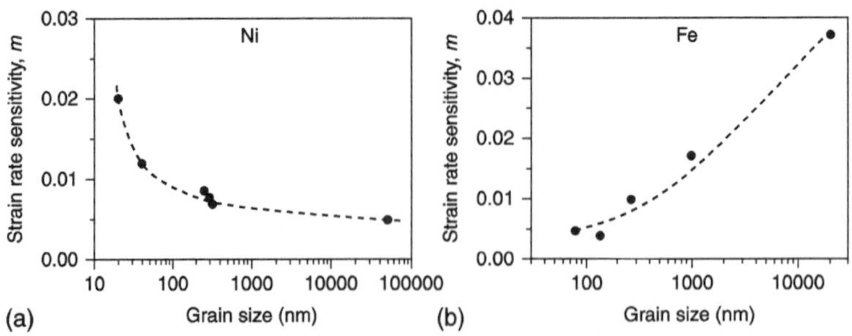

Source: The data are taken from Jia et al. (2003) [58] and Wei et al. (2006) [63].

where k is the Boltzmann constant and T is the absolute temperature of deformation. The activation volume decreases with decreasing grain size in fcc metals [53, 54]. The rate-controlling mechanism for conventional polycrystalline fcc metals is the glide movement of a dislocation through the dislocation forest. The activation area is equal to the spacing of forest dislocations multiplied by b and the activation volume is therefore equal to 100–1,000 × b^3 for dislocation mechanism. For UFG fcc materials, although the increase of the dislocation density yields a decrease in the average dislocation spacing, this trend is almost completely compensated by the increase of flow stress, therefore m only slightly increases. In the nanocrystalline regime, V^* decreases drastically to ~10 × b^3 as a consequence of the change in the rate-controlling mechanism [35]. Dislocation-boundary interactions or grain-boundary sliding [55] are two possible mechanisms operating in nanomaterials which may cause the reduction in V^* and the corresponding increase in m (see Figure 5.15a).

In bcc crystals, screw dislocation cores are dissociated into non-planar configurations [56], and the mobility of the edge components of dislocation loops is much higher than the mobility of the screw components [57]. For conventional polycrystalline bcc metals, the rate-controlling mechanism is the thermally activated motion of screw dislocations by the nucleation and propagation of kink pairs, which have an activation volume of only a few b^3 even for coarse-grained materials [58]. In bcc metallic materials V^* is not reduced considerably with decreasing grain size while the flow stress increases, which yields a continuous decrease of strain rate sensitivity as shown in Figure 5.15b for Fe. However, the constancy of V^* does not mean that the rate-controlling deformation mechanism is the same in the whole grain size regime since in bcc nanomaterials other mechanisms with a small activation volume (e.g. grain-boundary sliding) might become operative. The reduction of m for bcc metals with decreasing grain size contributes to the decrease in ductility at RT. The slight increase of m for UFG fcc materials is overwhelmed by the increase of flow stress, yielding a decrease of ductility with the reduction of grain size. At the same time, for nanocrystalline fcc metals, the large increment in m may cause a restoration of ductility with retaining the high yield strength. It is noted that the value of m for UFG and nanomaterials increases with decreasing strain rate and/ or increasing temperature of deformation [59].

Different strategies are developed to offset the low tensile ductility accompanying the high yield strength of nanocrystalline and UFG materials [59, 64]. For instance, the uniform incorporation of some micron-sized grains into the ultrafine-grained matrix yields an enhancement in the work hardening rate ($\frac{d\sigma}{d\varepsilon}$) because the dislocation density can increase considerably in the coarse grains during deformation [59, 65]. According to Equation 5.13, the higher work hardening rate may cause larger uniform elongation. The micron-sized grains, however, inevitably lower the yield strength of the otherwise nanocrystalline or ultrafine-grained material but the relative decrease of strength is much smaller than the relative increment in ductility. The bimodal grain-structure can be achieved by heterogeneous grain-growth during annealing of severely deformed materials such as Cu [66] or Ni [67]. For instance, Figure 5.16 compares the engineering stress–strain curves for UFG and bimodal Cu samples [59, 66]. The UFG Cu specimen is processed by eight passes of ECAP at RT and then rolled at liquid nitrogen temperature (cryorolling) for a reduction of the cross-sectional area of 93%. After ECAP the average grain size is 300 nm, which is refined to 200 nm due to

Figure 5.16 Room temperature tensile engineering stress–strain curves for Cu with different microstructures: UFG: processed by eight passes of ECAP at RT and then rolled at liquid nitrogen temperature for a reduction of the cross-sectional area of 93%; Bimodal: sample UFG annealed at 200 °C for 3 minutes; Coarse-grained: conventional sample. The strain rate for all the tests is 10^{-4} s^{-1}

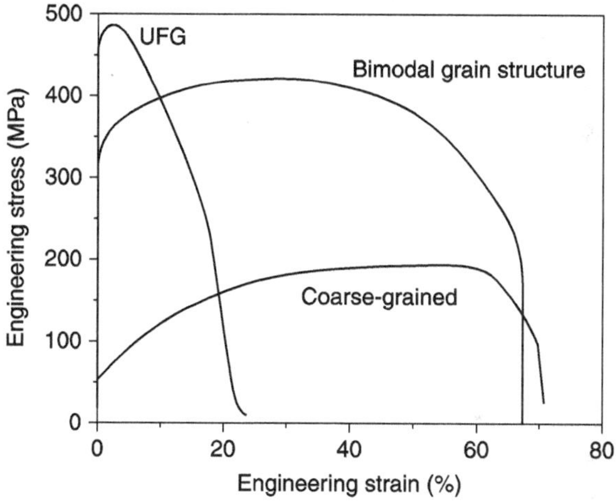

Source: The data are taken from Wang and Ma (2004) [59] and Wang et al. (2002) [66].

cryorolling. The annealing at 200°C for 3 minutes results in a partial recrystallisation of the cryorolled specimen, which gives a bimodal grain structure. The recrystallised grains with the size of ~2 μm are distributed homogeneously in the UFG matrix, having a mean grain size of 200 nm [59, 66]. In Figure 5.16 the deformation is uniform until the first derivative of the engineering strain remains non-negative. For the conventional coarse-grained material the uniform and total elongations are about 55% and 70%, respectively, which are reduced to 2% and 23% after SPD-processing, while there is a ten-fold increment in the yield strength.

Marked improvements in the uniform and total elongations to 30% and 67%, respectively, are achieved concurrent with pronounced strain hardening and with only a relatively small loss of strength in the annealed material with the bimodal microstructure as shown in Figure 5.16.

Another approach for retaining a good ductility in addition to high strength in pure UFG and nanomaterials is to synthesise samples with nanoscale growth twins. Nano-twinned 99.998% purity UFG Cu specimens were produced by pulsed electrodeposition technique from an electrolyte of $CuSO_4$ [68]. Due to their low energy, twin boundaries do not increase the stored energy very much, therefore from a thermodynamic point of view, it is beneficial to accommodate the orientation differences in the as-grown Cu by the formation of twin boundaries. It is noted that if a direct current is applied without pulses in electrodeposition, twins are scarcely observed in Cu. The twin boundary frequency is adjustable by modifying the electrodeposition parameters. When the deposition rate is high (for example, with a high current density or a large pH value of the solute) and/or the deposition temperature is low, more twins are formed. According to TEM investigations, the mean grain size of nano-twinned Cu processed by electrodeposition is ~500 nm, while the average twin boundary spacing is ~20 nm [68]. The sample has an (110) texture, which is usual for electrodeposited Cu specimens with growth twins. Although a coherent twin boundary has very low interface energy compared with a conventional grain boundary, it is also very effective in blocking dislocations. In the as-deposited sample, the dislocation density is very low. During deformation, dislocations are formed and accumulated at the twin boundaries. In the electrodeposited samples either with or without twins, the low initial dislocation density enables the increase of its value during plastic deformation, yielding a considerable work hardening that increases the uniform elongation compared with Cu specimens processed by SPD methods (see Table 5.2). The yield strength of the nano-twinned Cu is ~900 MPa, which is 2–3 times larger than the corresponding values for its counterparts with similar grain sizes since in the twinned material the characteristic length is rather the twin boundary spacing, which is much smaller than the grain size. Twin boundaries also act as dislocation sources. TEM investigations revealed that the high stress concentrations in the vicinity of steps and jogs along twin boundaries yielded nucleation of partial dislocations at these boundaries. This mechanism has much smaller activation volume, $V^* = 3\text{–}10 \times b^3$, than that of conventional full lattice dislocation motion (~100–1,000 × b^3) [54]. Equation 5.16 suggests that the strain-rate sensitivity at RT is determined by the product $V^*\sigma$. Although the flow stress, σ, is very high (~900 MPa) for nano-twinned Cu compared with UFG counterparts without twins (300–400 MPa), it cannot suppress the effect of the extremely low V^* in the product of $V^*\sigma$, which yields a relatively high strain-rate sensitivity, m ~ 0.036. Table 5.2 summarises the values of V^* and m for various Cu samples. For nano-twinned Cu, the strain rate sensitivity is 2–7 times larger than for the counterparts having similar grain size but no twin boundaries [30].

In alloys, nanoparticles and nano-precipitates pin dislocations that improve the dislocation storage capacity of the material. As a result, the work hardening rate increases, which leads to larger uniform elongation while elevating strength [35]. Applying SPD methods at high temperatures on supersaturated solid solutions or partially precipitated alloys, the large defect densities produced during processing act as nucleation sites of precipitates and therefore yield a large population of uniformly distributed nanoparticles [44]. Another strategy to increase the ductility in metallic materials

Table 5.2 The average grain size (d), the mean twin spacing (t), the activation volume (V^*), the strain rate sensitivity (m), the yield strength (σ_Y), the uniform and total elongations for various Cu samples at RT

Material	d [nm]	t[nm]	V^* [$\times b^3$]	m	σ_Y [MPa]	Uniform elongation [%]	Total elongation [%]
Coarse-grained Cu	40,000	–	1,000	0.006	25	55	70
Cu processed by 8 ECAP	200–300	–	48	0.02	400	1	14
Electrodeposited Cu (without pulses)	400–500	–	135	0.005	250	12	16
Electrodeposited Cu (with pulses)	400–500	20	12	0.036	900	12	14

Sources: The data are taken from Lu et al. (2005) [30] and Wang and Ma (2004) [59].

is hindering recovery during straining by reducing deformation temperature to cryogenic regime (liquid nitrogen temperature, ~77 K) and/or by increasing strain rate [35]. The reduced recovery of the dislocation structure increases the strain hardening rate and thereby the ductility is improved. For instance, for Cu processed by eight passes of ECAP, the uniform and total elongations increase from 1% to 13% and from 14% to 26%, respectively, when the temperature of tension is reduced from RT to 77 K at a strain rate of 10^{-4} s^{-1} [59]. For this material, the increment of strain rate from 10^{-4} s^{-1} to 10^{-1} s^{-1} at RT yields an improvement in uniform and total elongations from 1% to 3% and from 14% to 19%, respectively. At cryogenic temperature, the uniform elongation changes from 13% to 19%, while the total elongation increases from 26% to 40% when the strain rate is raised from 10^{-4} s^{-1} to 10^{-1} s^{-1} [59].

SPD methods, such as ECAP, may also yield a combination of good ductility and high strength, if SPD-processing is performed beyond a certain level of strain [69]. For instance, the total elongation of Cu processed by ECAP at RT first decreases with increasing number of passes, but after 16 passes the elongation is restored to ~55%, which is very close to the value obtained on coarse-grained material (60–70%) while the yield strength remains high (~400 MPa). After 4–6 passes of ECAP, the grain size in Cu saturates at a value of 100–200 nm, but the microstructure development continues, and the fraction of high angle grain boundaries increases [69, 70] and/or the dislocation density slightly decreases [31, 71]. The latter structural recovery at large strains is accompanied by a decrease in the grain boundary thickness, corresponding to an evolution from non-equilibrium boundaries to a more equilibrated structure [72, 73]. The larger fraction of high-angle grain boundaries facilitates grain-boundary sliding as a mechanism of plastic deformation [69]. Deformation entirely by grain-boundary sliding would ideally result in a strain rate sensitivity of 0.5, whereas conventional dislocation slip typically results in strain rate sensitivities less than 0.05. The relatively high value of $m = 0.14$ obtained for Cu processed by 16 passes of ECAP suggests that a mixture of grain-boundary sliding and dislocation slip occurs in nanostructured metals processed by SPD for large strains [69]. The higher strain rate sensitivity improves the ductility without sacrificing the large yield strength.

5.5 Influence of sintering conditions on strength and ductility of consolidated nanomaterials

In section 4.3, the defect structures in UFG Ni samples sintered from nanopowders under different consolidation conditions were reviewed. In this section, the mechanical behaviours of these materials are compared. The samples are consolidated from nanopowders with the particle size of 100 or 50 nm. The sintering is carried out by HIP or SPS method. The details of the consolidation procedures are given in section 4.3 and also listed in Table 5.3. Both the nanopowders and the consolidated materials contain NiO. The amount of NiO is characterised by the integrated intensity ratio of the NiO and Ni peaks on the x-ray diffractograms at $2\Theta = 37.4°$ and 44.6°, respectively. The intensity ratios for the initial powders having particle sizes of 100 and 50 nm are 0.5% and 1.2 %, respectively, which does not change during HIP-processing as a result of the careful isolation of the sample from air. On the other hand, in SPS-processing, the capsule containing the powder was broken in air, which yields an

Table 5.3 The relative density, the intensity ratio (I_{NiO}/I_{Ni}) of the x-ray peaks for NiO and Ni at $2\Theta = 37.4°$ and $44.6°$, respectively, the mean grain size (*d*) obtained from TEM images, the yield strength, maximum strength and the strain to failure determined by compression for Ni samples processed from nanopowders by HIP or SPS [62, 74]

Powder particle size [nm]	Consolidation conditions	Relative density [%]	I_{NiO}/I_{Ni} [%]	*d* [nm]	Yield strength [MPa]	Maximum strength [MPa]	Strain to failure
100	HIP, 140 MPa, 700°C, 150 min, in Ar	95.5	0.5±0.1	403	542	789	0.35
100	SPS, 150 MPa, 500°C, 1 min, in air	95.4	0.9±0.1	294	682	873	0.21
100	SPS, 100 MPa, 500°C, 4 min, in Ar	94.0	0.6±0.1	306	660	878	0.33
50	SPS, 150 MPa, 500°C, 1 min, in air	94.5	1.5±0.1	250	1,022	1,178	0.04

increase of NiO content from 0.5% to 0.9% and from 1.2% to 1.5% for the powders with particle sizes of 100 and 50 nm, respectively [74]. When the consolidation by SPS is performed in Ar, the amount of NiO does not increase significantly (see Table 5.3). The grain size and the defect densities are listed in Table 4.2 in Chapter 4 for the samples sintered by SPS in air and by HIP in Ar.

Figure 5.17 shows the stress–plastic strain curves obtained for the four consolidated Ni samples by quasi-static compression at room temperature. The yield strength, the maximum strength and the strain to failure are listed in Table 5.3. The yield strength data are plotted as a function of the inverse square-root of the grain size in Figure 5.18 (Hall-Petch plot). In the figure, the values obtained on the four UFG Ni samples are completed by the strength (96 MPa) determined on a coarse-grained (grain size is ~5 μm)

Figure 5.17 Compression stress–strain curves for Ni samples consolidated from powders with the grain size of 50 or 100nm, in Ar or air and by HIP or SPS

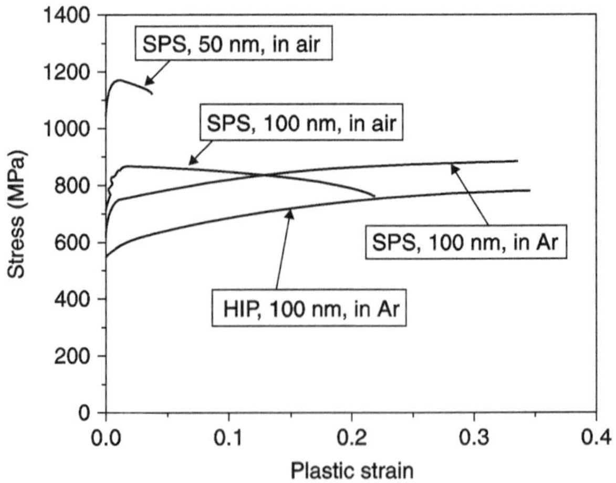

Source: After Bui et al. (2010) [62] with permission from Elsevier.

specimen consolidated by SPS. In addition to the strength values obtained on the samples processed by powder metallurgy (denoted by solid squares), data determined on oxide-free Ni specimens produced by electrodeposition (marked by open circles) are also plotted in Figure 5.18 [62]. The latter values give a Hall-Petch relationship with the values of σ_0 = 37 MPa and k = 5,538 MPa·nm$^{1/2}$ represented by a solid straight line in Figure 5.18. The yield strength data of the UFG samples processed by powder metallurgy are higher than the values predicted from the Hall-Petch relationship since not only the grain size but also the NiO

Figure 5.18 **Hall-Petch relation between the yield strength (σ_Y) and the grain size (d) for Ni samples processed by electrodeposition and powder metallurgy. The solid line represents a linear fit on the data obtained for electrodeposited Ni specimens**

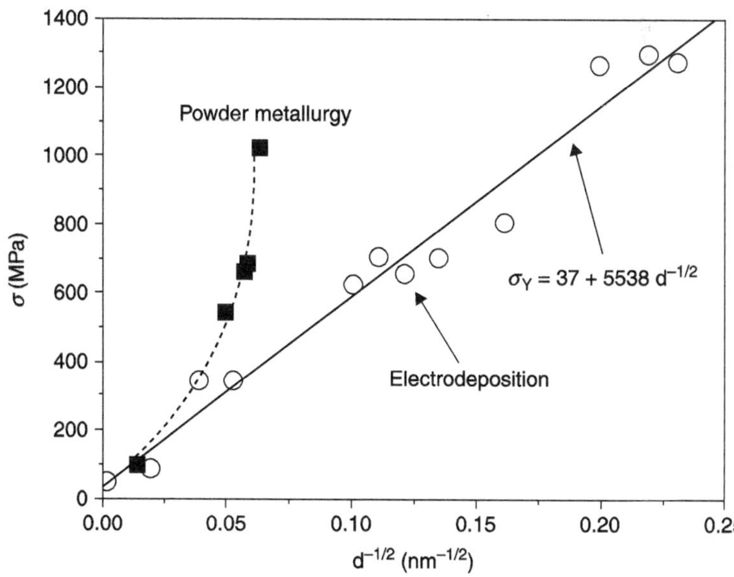

Source: After Bui et al. (2010) [62] with permission from Elsevier.

dispersoids contribute to the strength. In the case of the coarse-grained Ni specimen consolidated by SPS, the oxide content is marginal, therefore its strength is determined basically by the grain size as shown in Figure 5.18. Assuming a simple linear addition of the strength contributions from the grain size and the oxide particles, the oxide strengthening in the UFG samples processed by powder metallurgy can be estimated by subtracting the strength contribution of the grain size from the experimental yield strength. The strength caused by the grain size can be calculated from the Hall-Petch relationship obtained for the oxide-free electrodeposited Ni samples. Figure 5.19 shows that the calculated strength contribution of NiO dispersoids correlates well with the intensity ratio I_{NiO}/I_{Ni}, i.e. with the oxide volume fraction. Figure 5.18 suggests that the smaller the grain size, the higher the strength contribution of NiO. The main reason of this behaviour is the fact that the majority of oxide in the initial

Figure 5.19 The strength contribution of NiO dispersoids in sintered Ni samples as a function of the intensity ratio (I_{NiO}/I_{Ni}) of the x-ray diffraction peaks for NiO and Ni at $2\Theta=37.4°$ and $44.6°$, respectively

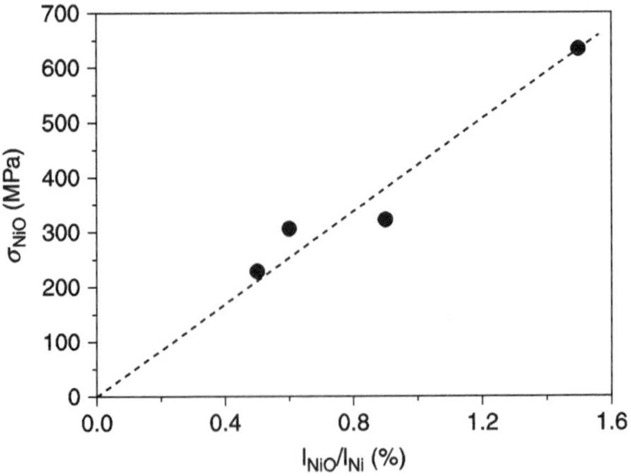

powder can be found on the particles' surfaces, therefore the smaller powder particle size is accompanied by higher oxide content. As the smaller particle size usually yields smaller grain size after consolidation, the NiO content and its strengthening contribution increase with decreasing grain size. It should be noted that in addition to the initial particle size, the consolidation atmosphere also influences the oxide content in the sintered material as revealed in Table 5.3. This effect will be discussed in the following paragraphs.

Figure 5.17 shows that the UFG Ni samples processed from the powder having a mean particle size of 100 nm by HIP or SPS have very different mechanical behaviours. The specimen processed by SPS has higher yield strength (682 MPa) than that obtained for the sample consolidated by HIP method (542 MPa) as a result of the higher oxide dispersoid content and the smaller grain size of the former specimen. The smaller grain size is due to the reduced temperature and time of SPS procedure, while the higher NiO content is a result of the processing of the powder in air. Comparing the mechanical behaviours of the two samples, it is revealed that the HIP-processed sample hardens in the whole range of strain while in the case of SPS-processed specimen the flow stress saturates at a small strain value ($\varepsilon \approx 0.04$) and after a very short plateau the sample softens until rupture. The values of strain to failure are 0.22 and 0.35 for the SPS- and HIP-processed samples, respectively [74].

For a better understanding of the difference in the mechanical behaviours of the HIP- and SPS-processed samples, the defect structures were determined after the compression test by x-ray line profile analysis. Figures 5.20a and 5.20b show the dislocation density and the twin boundary frequency for the HIP- and SPS-processed samples before and after compression. It can be seen that during room temperature deformation the dislocation density increased from 5.6 ± 0.5 to $16 \pm 2 \times 10^{14}$ m^{-2} for the HIP-processed sample and from 5.1 ± 0.5 to $21 \pm 3 \times 10^{14}$ m^{-2} in the case of the SPS-processed specimen. At the same time, the twin boundary frequency decreased during compression for both samples: in the case of the HIP-processed sample from $0.32 \pm 0.03\%$ to $0.02 \pm 0.02\%$ and for the SPS-processed specimen from $0.29 \pm 0.03\%$ to $0.18 \pm 0.02\%$. Figures 5.20c and 5.20d show the distribution of the angle of misorientation

Figure 5.20 (a) The dislocation density and (b) the twin boundary frequency for the HIP- and SPS-processed samples before and after compression test as determined by x-ray line profile analysis. The distribution of the angle of misorientation between the grains in the (c) HIP-processed sample and after compression up to 10% strain and (d) in the SPS-processed sample and after compression up to 10% strain as determined by EBSD

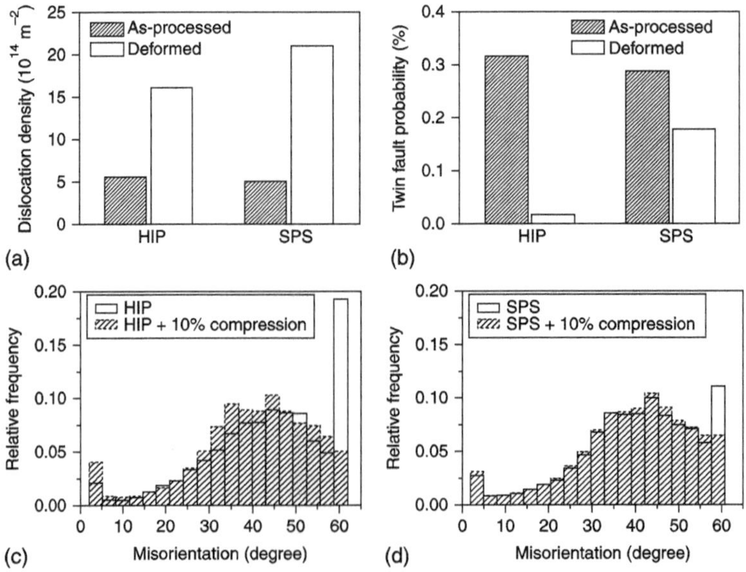

Source: Reproduced from Gubicza et al. (2009) [74] with permission from Cambridge University Press.

between the grains in the as-consolidated samples and in the specimens compressed up to 10% strain as determined by electron backscatter diffraction (EBSD) experiments. For the as-consolidated samples there is a peak at about 60° in the misorientation distribution, which corresponds to twin boundaries [67, 75]. After compression at 10% strain, this peak disappeared for both samples, indicating the reduction of densities of twin boundaries. This observation is in agreement with the results of x-ray line profile analysis.

The increase of the dislocation density and the decay of twins (referred to as untwinning) are the result of the interaction between dislocations and twin boundaries during compression. This interaction first yields steps on twin boundaries as indicated by black arrows in the TEM image of Figure 5.21a [74]. The mechanism of step-formation is depicted in Figure 5.21b. Twin boundaries act as obstacles to dislocation gliding and the blocked dislocations may dissociate [67]. For example, if a dislocation having a Burgers vector of $\frac{1}{2}[101]$ and gliding on plane $(\bar{1}11)$ reaches a twin boundary on plane (111), this dislocation may dissociate into a Shockley-partial and a non-glissile Frank dislocation, such as $\frac{1}{2}[101] \rightarrow \frac{1}{6}[1\bar{2}1] + \frac{1}{3}[111]$. The Shockley-partial with the Burgers-vector of $\frac{1}{6}[1\bar{2}1]$ may slip on plane (111) resulting in a narrowing of a part of the twin by one (111) lattice spacing. If a slip band intersects a twin boundary on plane (111), the slipping of Shockley-partials on successive planes (111) yields a partial untwinning, as discussed in Chapter 3.

It is a general observation for both HIP- and SPS-processed samples that untwinning can be observed only in large grains having grain sizes higher than about 500 nm because in larger grains the contribution of dislocations to plastic deformation is higher. In Chapter 4, Figures 4.9c and 4.9d

Figure 5.21 **(a) TEM image showing twins in a large grain after compression test of Ni sample processed from a powder with particle size of 100nm by SPS in air. The arrows show steps on twin boundaries as a result of untwinning process. The untwinning mechanism is illustrated in (b) showing the Burgers-vectors of dislocations interacting with a twin boundary (TB)**

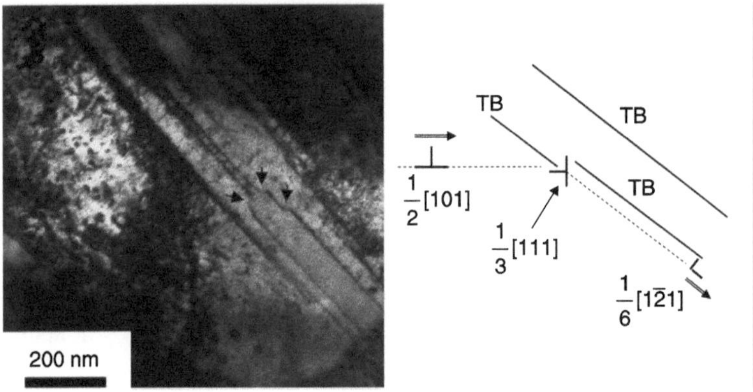

(a) (b)

Source: Reproduced from Gubicza et al. (2009) [74] with permission from Cambridge University Press.

show that the size distribution of grains containing twins follows the total grain size distribution in both consolidated Ni samples. As the HIP-processed sample has larger grains and untwinning mainly occurred in large grains, the decrease of twin probability during compression is expected to be higher for the HIP-processed sample. This is in agreement with the experimental results of x-ray line profile analysis, which shows that the reduction of twin probability is much higher for the HIP-processed sample than in the case of the SPS-processed specimen (see Figure 5.20b). The larger strain to failure for the HIP-processed sample may also have a contribution to the higher degree of untwinning in this specimen. It is noted that in the case of the sample produced from the powder having a mean particle size of 50 nm by SPS, the dislocation density increases from 3.7 to 8.5×10^{14} m^{-2} and the twin probability decreases from 0.25% to 0.20% in the compression test. The smaller changes of defect densities in this case can be explained by the very small value of strain to failure.

In the beginning of compression test the strain hardening for the SPS-processed sample is larger and the flow stress saturates at a relatively low strain value. This can be explained by the larger fraction of oxide dispersoids which interact with the dislocations emitted from grain boundaries leaving dislocation loops around the NiO particles, thereby increasing the dislocation density very quickly [76]. It is noted that the low strain value at saturation does not exclude a high value of the saturated dislocation density, as can be seen in the case of the sample processed by SPS ($21 \pm 3 \times 10^{14}$ m^{-2}). After saturation of the flow stress, strain softening occurs during deformation of the SPS-processed sample while the HIP-processed specimen hardens until failure (see Figure 5.17). Figure 5.22 shows scanning electron microscopy (SEM) images taken on the surface of the SPS-processed sample deformed by compression to the strain value of 7% (a), 18% (b), 22% (c) and for the HIP-processed specimen at a strain of 25% (d). The loading axis is vertical on the figure. It can be seen that cracks (indicated by arrows) form in the SPS-processed sample during compression and the concentration and length of cracks increase with increasing strain. At the same time, no decohesion or crack formation was observed in the case of the HIP-processed sample compressed up to 25% strain. TEM pictures reveal that the cracks are formed at the grain boundaries of the SPS-processed sample during compression tests. The crack formation is most probably the reason for the flow softening

Figure 5.22 SEM images showing the surface of the sample processed by SPS in air deformed by compression to the strain values of (a) 7%, (b) 18%, and (c) 22% and (d) the surface of the HIP-processed specimen at a strain of 25%. The loading axis is vertical in the figures. Some cracks are indicated by black arrows

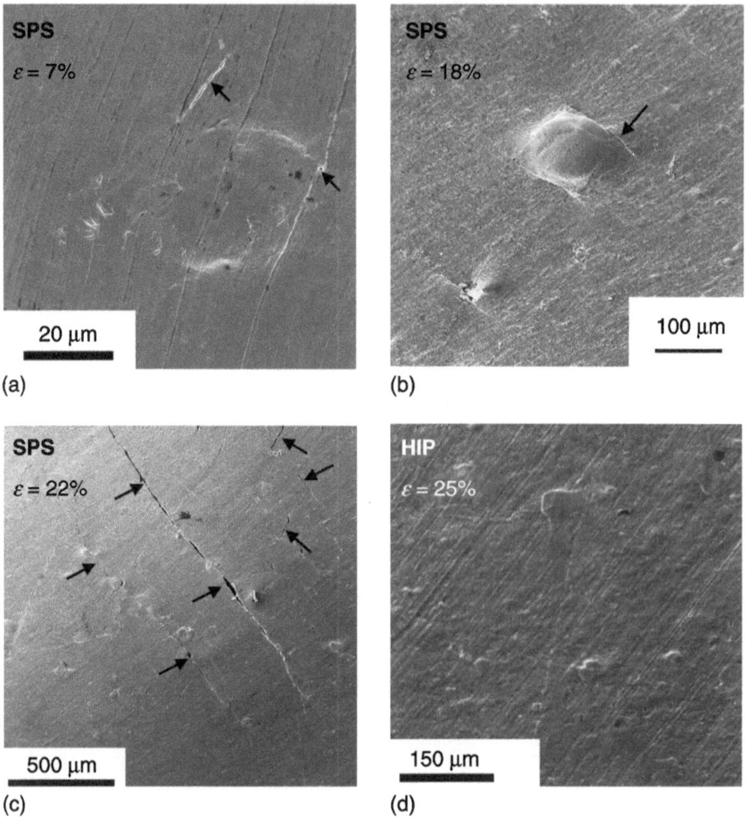

Source: Reproduced from Gubicza et al. (2009) [74] with permission from Cambridge University Press.

during compression test and the smaller strain to failure in the case of the SPS-processed sample. The crack formation can be attributed to the higher flow stress level during deformation of the SPS-processed sample (see Figure 5.17), which results in a larger probability of debonding between the grain boundaries. Additionally, the higher total oxide content in the SPS-processed sample probably involves higher oxide content on the grain boundaries, which may reduce the grain boundary strength, resulting in an easier cracking during deformation [77].

Mechanical behaviour of materials sintered from blends of powders with different grain sizes

In section 4.4, the grain size distribution and the defect structure were studied in Ni materials consolidated from blends of nanopowder (NP) and coarse-powder (CP). In this section, the mechanical properties of these materials are studied. In addition to the bulk samples processed from 100% NP and 100% CP powders (referred to as samples UFG and CG, respectively), two additional specimens with different volume fractions of powders NP and CP were prepared. The volume fractions of coarse-grained (CG) and ultrafine-grained (UFG) components are 61% and 39% for sample A, respectively, while these values are 39% and 61% for sample B [78]. Figure 5.23a shows the resulting true stress–true strain curves for samples CG, A, B and UFG. The yield strength was plotted as a function of the UFG volume fraction in Figure 5.23b. It can be seen that the yield strength increases with increasing the amount of the UFG content of the samples. However, the yield strength values determined for samples A and B can be found under the straight line connecting the datum points of the fully CG and UFG samples. This means that the yield strength of the composite-like materials cannot be obtained by a linear interpolation from the strength values of the fully CG and UFG specimens [78].

Figure 5.23 (a) True stress–logarithmic plastic strain curves obtained by compression test at RT for samples CG, A, B and UFG consolidated from blends of nano- and coarse-powders with different volume fractions. The volume fraction of the CG component is also indicated. (b) The yield strength values obtained from the stress–strain curves in (a)

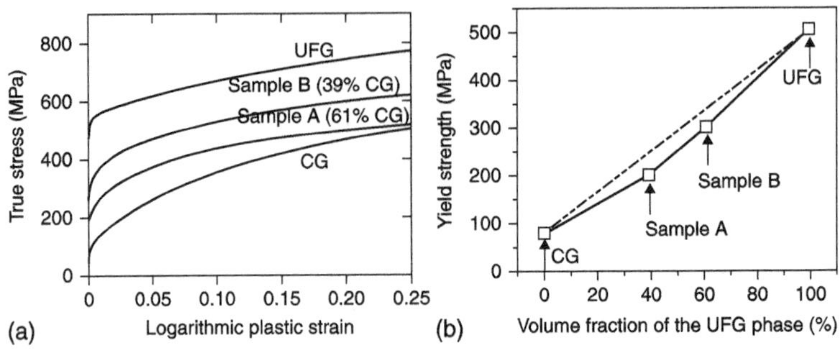

Source: Reprinted from Dirras et al. (2010) [78] with permission from Elsevier.

The strength of the CG and UFG components are characterised locally by nanoindentation. The nanohardness distributions for the fully CG specimen, samples A and B, and for the fully UFG specimen are shown in Figure 5.24. The mean hardness increases in the following order: specimen CG (2.4 ± 0.1 GPa), sample A (3.3 ± 0.1 GPa), sample B (4.9 ± 0.1 GPa) and specimen UFG (7.1 ± 0.1 GPa) [78]. It can be seen that the mean nanohardness increases with increasing the UFG content of the samples, which is in line with the trend observed for the yield strength. At the same time, the hardness for samples A and B is smaller than the values interpolated from the hardness values characteristic for the fully CG and UFG specimens (see Figure 5.25). This observation is in line with the results obtained for the yield strength.

It can be seen from Figure 5.24 that the hardness distributions for samples A and B are asymmetric, having

Figure 5.24 **The nanohardness distributions for (a) specimen CG, (b) sample A, (c) sample B and (d) specimen UFG**

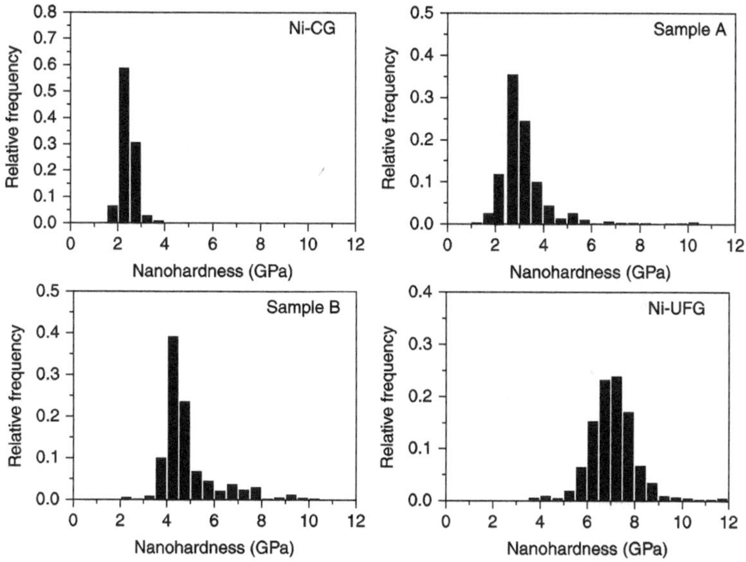

Source: Reprinted from Dirras et al. (2010) [78] with permission from Elsevier.

long tail in the higher hardness region. This tail part is most probably related to the UFG component of the microstructure. The low-hardness part of the distribution is assumed to be related to the CG component. The mean hardness of the CG component is determined by averaging the lowest hardness values up to 61% and 39% of the total number of indentations for samples A and B, respectively (these are the relative fractions of the CG components). The mean hardness of the UFG component is estimated by averaging 39% and 61% of hardness numbers having the highest values for samples A and B, respectively. The hardness values of the CG and UFG components for different samples are plotted in Figure 5.25. The hardness of the fully CG sample is lower than that for the CG fraction in sample A, which is lower than the hardness

Figure 5.25 The nanohardness of the CG and UFG fractions, and their volume-weighted average as a function of the volume fraction of the UFG component for the as-processed specimen CG, sample A, sample B and specimen UFG. The dashed line represents the linear interpolation between the values characteristic for the fully CG and UFG samples

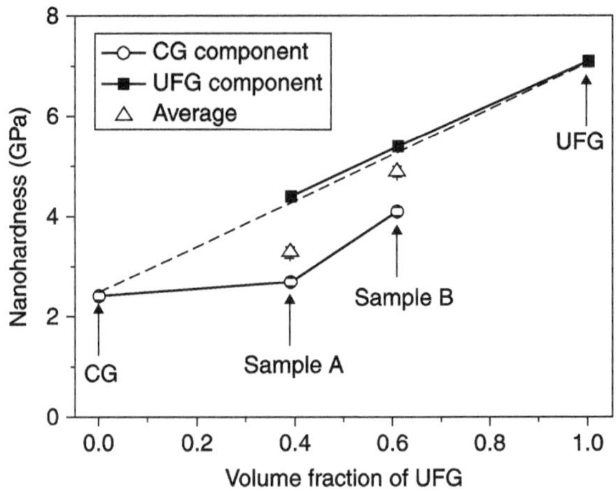

Source: Reprinted from Dirras et al. (2010) [78] with permission from Elsevier.

of the CG component in sample B [78]. This can be explained by the decrease of the grain size in the CG component with increasing UFG volume fraction as discussed in detail in section 4.4 ('barrier effect'). It should be noted that the relative number of indents located in the CG grains may differ slightly from the volume fraction of the CG component, which may result in slight deviation of the calculated mean hardness from the real value. It should also be noted that some hardness values measured in the CG grains may be affected by the hardness of the UFG grains, and vice versa, as the size of an indent is determined by the strength of a volume around the indent having a radius of about 20 times the penetration depth (this depth was 100–250 nm). For example, close to the interfaces between the CG and UFG grains the hardness of the CG grains may be higher than in the centre of the CG grains, or the hardness of the UFG component may be lower if there is a CG grain under the measured UFG region.

Figure 5.25 shows that in the composite samples the hardness of the UFG component is lower than that in the fully UFG sample, and it seems that the higher the CG fraction, the lower the hardness of the UFG component [78]. This can be explained by the 'shielding effect' of the CG grains on the UFG volumes during sintering as discussed in section 4.4. As a result of this effect, the dislocation density is smaller and therefore the hardness is lower in the UFG components of samples A and B than in the fully UFG specimen. The 'shielding effect' of the CG component may also yield incomplete particle bonding within the UFG component, which also contributes to the lower hardness. In the case of sample B, the lower hardness of the UFG component may be partially caused by the lower relative mass density, which is most probably associated with a higher porosity due to the lower temperature and shorter duration of sintering (see section 4.4). The smaller hardness of the UFG component in samples A and B compared with the fully UFG specimen is primarily responsible for the smaller hardness and yield strength of the composite samples than the values interpolated from the data characteristic for the fully CG and UFG samples (see Figures 5.23b and 5.25).

5.7 References

1. V. Yamakov, D. Wolf, M. Salazar, S.R. Phillpot and H. Gleiter, 'Length-scale effects in the nucleation of extended

dislocations in nanocrystalline Al by molecular-dynamics simulation', *Acta Materialia*, 49 (2001) 2713–2722.

2. H. Van Swygenhoven, 'Grain boundaries and dislocations', *Science*, 296 (2002) 66–67.
3. V. Yamakov, D. Wolf, S.R. Phillpot, A.K. Mukherjee and H. Gleiter, 'Dislocation processes in the deformation of nanocrystalline aluminium by molecular-dynamics simulation', *Nature Materials*, 1 (2002) 1–4.
4. X.Z. Liao, Y.H. Zhao, S.G. Srinivasan, Y.T. Zhu, R.Z. Valiev and D.V. Gunderov, 'Deformation twinning in nanocrystalline copper at room temperature and low strain rate', *Applied Physics Letters*, 84 (2004) 592–594.
5. X.Z. Liao, F. Zhou, E.J. Lavernia, S.G. Srinivasan, M.I. Baskes, D.W. He and Y.T. Zhu, 'Deformation mechanism in nanocrystalline Al: Partial dislocation slip', *Applied Physics Letters*, 83 (2003) 632–634.
6. Y.T. Zhu and X.Z. Liao, 'Nanostructured metals: Retaining ductility', *Nature Materials*, 3 (2004) 351–352.
7. A.G. Froseth, P.M. Derlet and H. Van Swygenhoven, 'Dislocations emitted from nanocrystalline grain boundaries: nucleation and splitting distance', *Acta Materialia*, 52 (2004) 5863–5870.
8. Y.T. Zhu, X.Z. Liao, S.G. Srinivasan, Y.H. Zhao and M.I. Baskes, 'Formation mechanism of wide stacking faults in nanocrystalline Al', *Applied Physics Letters*, 85 (2004) 5049–5051.
9. Y.T. Zhu, X.Z. Liao and X.L. Wu, 'Deformation twinning in nanocrystalline materials', *Progress in Materials Science*, 57 (2012) 1–62.
10. L. Balogh, R.B. Figueiredo, T. Ungár and T.G. Langdon, 'The contributions of grain size, dislocation density and twinning to the strength of a magnesium alloy processed by ECAP', *Materials Science and Engineering*, A 528 (2010) 533–538.

11. A. Galiyev, R. Kaibyshev and G. Gottstein, 'Correlation of plastic deformation and dynamic recrystallization in magnesium alloy ZK60', *Acta Materialia*, 49 (2001) 1199–1207.
12. X.L. Wu, K.M. Youssef, C.C. Koch, S.N. Mathaudhu, L.J. Kecskés and Y.T. Zhu, 'Deformation twinning in a nanocrystalline hcp Mg alloy', *Scripta Materialia*, 64 (2011) 213–216.
13. M.F. Ashby and R.A. Verall, 'Diffusion-accommodated flow and superplasticity', *Acta Metallurgica*, 21 (1973) 149–163.
14. P. Yavari and T.G. Langdon, 'Cavitation in high purity aluminium during fatigue at elevated temperatures', *Journal of Materials Science Letters*, 2 (1983) 522–524.
15. N.Q. Chinh, P. Szommer, Z. Horita and T.G. Langdon, 'Experimental Evidence for Grain-Boundary Sliding in Ultrafine-Grained Aluminum Processed by Severe Plastic Deformation', *Advanced Materials*, 18 (2006) 34–39.
16. N.Q. Chinh, J. Gubicza and T.G. Langdon, 'Characteristics of face-centered cubic metals processed by equal-channel angular pressing', *Journal of Materials Science*, 42 (2007) 1594–1605.
17. A. Vinogradov, S. Hashimoto, V. Patlan and K. Kitagawa, 'Atomic force microscopic study on surface morphology of ultra-fine grained materials after tensile testing', *Materials Science and Engineering*, A 319–321 (2001) 862.
18. Y. Huang and T.G. Langdon, 'Using atomic force microscopy to evaluate the development of mesoscopic shear planes in materials processed by severe plastic deformation', *Materials Science and Engineering*, A 358 (2003) 114–121.
19. C.Y. Yu, P.L. Sun, P.W. Kao and C.P. Chang, 'Mechanical properties of submicron-grained aluminum', *Scripta Materialia*, 52 (2005) 359–365.

20. H. Van Swygenhoven and A. Caro, 'Plastic behavior of nanophase Ni: A molecular dynamics computer simulation', *Applied Physics Letters*, 71 (1997) 1652.
21. J. Schiotz, F.D. Di Tolla and K.W. Jacobsen, 'Softening of nanocrystalline metals at very small grain sizes', *Nature*, 391 (1998) 561–563.
22. D. Jia, Y.M. Wang, K.T. Ramesh, E. Ma, Y.T. Zhu and R.Z. Valiev 'Deformation behavior and plastic instabilities of ultrafine-grained titanium', *Applied Physics Letters*, 79 (2001) 611–613.
23. M.A. Meyers, A. Mishra and D.J. Benson, 'Mechanical properties of nanocrystalline materials', *Progress in Materials Science*, 51 (2006) 427–556.
24. E.O. Hall, 'The deformation and ageing of mild steel: III discussion of results', *Proceedings of the Physical Society London*, B64 (1951) 747–753.
25. N.J. Petch, 'The cleavage strength of polycrystals', *Journal of the Iron and Steel Institute*, 174 (1953) 25–28.
26. N. Hansen and R. Ralph, 'The strain and grain size dependence of the flow stress of copper', *Acta Metallurgica*, 30 (1982) 411–417.
27. A.H. Chokshi, A. Rosen, J. Karch and H. Gleiter 'On the validity of the Hall-Petch relationship in nanocrystalline materials', *Scripta Metallurgica*, 23 (1989) 1679–1684.
28. G.E. Fougere, J.R. Weertman, R.W. Siegel and S. Kim, 'Grain-size dependent hardening and softening of nanocrystalline Cu and Pd', *Scripta Metallurgica et Materialia*, 26 (1992) 1879–1883.
29. P.G. Sanders, C.J. Youngdahl and J.R. Weertman, 'The strength of nanocrystalline metals with and without flaws', *Materials Science and Engineering*, A 234–236 (1997) 77–82.

30. L. Lu, R. Schwaiger, Z.W. Shan, M. Dao, K. Lu and S. Suresh, 'Nano-sized twins induce high rate sensitivity of flow stress in pure copper', *Acta Materialia*, 53 (2005) 2169–2179.
31. J. Gubicza, N.Q. Chinh, J.L. Lábár, S. Dobatkin, Z. Hegedűs and T.G. Langdon, 'Correlation between microstructure and mechanical properties of severely deformed metals', *Journal of Alloys and Compounds*, 483 (2009) 271–274.
32. J. Lian, B. Baudelet and A.A. Nazarov, 'Model for the prediction of the mechanical behaviour of nanocrystalline materials', *Materials Science and Engineering*, A 172 (1993) 23–29.
33. Y. Estrin, H.S. Kim and F.R.N. Nabarro, 'A comment on the role of Frank-Read sources in plasticity of nanomaterials', *Acta Materialia*, 55 (2007) 6401–6407.
34. T. Nieh and J. Wadsworth, 'High-strain-rate superplasticity in aluminum matrix composites', *Scripta Metallurgica*, et Materialia 25 (1991) 955–958.
35. M.A. Meyers, A. Mishra and D.J. Benson, 'The deformation physics of nanocrystalline metals: experiments, analysis, and computations', *Journal of the Minerals, Metals and Materials Society*, 58 (2006) 41–48.
36. C.S. Pande and K.P. Cooper, 'Nanomechanics of Hall-Petch relationship in nanocrystalline materials', *Progress in Materials Science*,. 54 (2009) 689–706.
37. J. Schiotz and K.W. Jacobsen, 'A maximum in the strength of nanocrystalline copper', *Science*, 301 (2003) 1357–1359.
38. J. Gubicza, N.Q. Chinh, Z. Horita and T.G. Langdon, 'Effect of Mg addition on microstructure and mechanical properties of aluminum', *Materials Science and Engineering*, A 387–389 (2004) 55–59.

39. D.A. Hughes and N. Hansen, 'Microstructure and strength of nickel at large strains', *Acta Materialia*, 48 (2000) 2985–3004.
40. J. Gubicza, N.Q. Chinh, Gy. Krállics, I. Schiller and T. Ungár, 'Microstructure of ultrafine-grained fcc metals produced by severe plastic deformation', *Current Applied Physics*, 6 (2006) 194–199.
41. J. Gubicza, S.V. Dobatkin, E. Khosravi, A.A. Kuznetsov and J.L. Lábár, 'Microstructural stability of Cu processed by different routes of severe plastic deformation', *Materials Science and Engineering*, A 528 (2011) 1828–1832.
42. J. Gubicza, N.Q. Chinh, J.L. Lábár, Z. Hegedűs, C. Xu and T.G. Langdon, 'Microstructure and yield strength of severely deformed silver', *Scripta Materialia*, 58 (2008) 775–778.
43. L. Balogh, T. Ungár, Y. Zhao, Y.T. Zhu, Z. Horita, C. Xu and T.G. Langdon, 'Influence of stacking-fault energy on microstructural characteristics of ultrafine-grain copper and copper-zinc alloys', *Acta Materialia*, 56 (2008) 809–820.
44. J. Gubicza, I. Schiller, N.Q. Chinh, J. Illy, Z. Horita and T.G. Langdon, 'The effect of severe plastic deformation on precipitation in supersaturated Al-Zn-Mg alloys', *Materials Science and Engineering*, A 460–461 (2007) 77–85.
45. F. Hernandez Olivares and J. Gil Sevillano, 'A quantitative assessment of forest-hardening in FCC metals', *Acta Metallurgica*, 35 (1987) 631–641.
46. E. Schafler, K. Simon, S. Bernstorff, P. Hanák, G. Tichy, T. Ungár and M.J. Zehetbauer, 'A second-order phase-transformation of the dislocation structure during plastic deformation determined by in situ synchrotron X-ray diffraction', *Acta Materialia*, 53 (2005) 315–322.

47. U.F. Kocks, 'A statistical theory of flow stress and work-hardening', *Philosophical Magazine*, 13 (1966) 541–566.
48. N.Q. Chinh, G. Horváth, Z. Horita and T.G. Langdon, 'A new constitutive relationship for the homogeneous deformation of metals over a wide range of strain', *Acta Materialia*, 52 (2004) 3555–3563.
49. N. Tsuji, Y. Ito, Y. Saito and Y. Minamino, 'Strength and ductility of ultrafine grained aluminum and iron produced by ARB and annealing', *Scripta Materialia*, 47 (2002) 893–899.
50. C. Xu, W. Dixon, M. Furukawa, Z. Horita and T.G. Langdon, 'Developing superplasticity in a spray-cast aluminum 7034 alloy through equal-channel angular pressing', *Materials Letters*, 57 (2003) 3588–3592.
51. C. Xu, M. Furukawa, Z. Horita and T.G. Langdon, 'Using ECAP to achieve grain refinement, precipitate fragmentation and high strain rate superplasticity in a spray-cast aluminum alloy', *Acta Materialia*, 51 (2003) 6139–6149.
52. C. Xu, M. Furukawa, Z. Horita and T.G. Langdon, 'The evolution of homogeneity and grain refinement during equal-channel angular pressing: A model for grain refinement in ECAP', *Acta Materialia*, 53 (2005) 749–758.
53. M. Dao, L. Lu, R.J. Asaro, J.T.M. De Hosson and E. Ma, 'Toward a quantitative understanding of mechanical behavior of nanocrystalline metals', *Acta Materialia*, 55 (2007) 4041–4065.
54. R.J. Asaro and S. Suresh, 'Mechanistic models for the activation volume and rate sensitivity in metals with nanocrystalline grains and nano-scale twins', *Acta Materialia*, 53 (2005) 3369–3382.

55. H. Conrad, 'Grain-size dependence of the flow stress of Cu from millimeters to nanometers', *Metallurgical and Materials Transactions*, A 35A (2004) 2681–2695.
56. M.S. Duesbery and V. Vitek, 'Plastic anisotropy in bcc transition metals', *Acta Materialia*, 46 (1998) 1481–1492.
57. A.H.W. Ngan and M. Wen, 'Atomistic simulations of Paidar–Pope–Vitek lock formation in Ni_3Al', *Computational Materials Science*, 23 (2002) 139–145.
58. D. Jia, K.T. Ramesh and E. Ma, 'Effects of nanocrystalline and ultrafine grain sizes on constitutive behavior and shear bands in iron', *Acta Materialia*, 51 (2003) 3495–3509.
59. Y.M. Wang and E. Ma, 'Three strategies to achieve uniform tensile deformation in a nanostructured metal', *Acta Materialia*, 52 (2004) 1699–1709.
60. F. Dalla Torre, H. Van Swygenhoven and M. Victoria, 'Nanocrystalline electrodeposited Ni: microstructure and tensile properties', *Acta Materialia*, 50 (2002) 3957–3970.
61. C. Gu, J. Lian, Z. Jiang and Q. Jiang, 'High corrosion-resistance nanocrystalline Ni coating on AZ91D magnesium alloy', *Scripta Materialia*, 54 (2006) 579–584.
62. Q.H. Bui, G. Dirras, S. Ramtani and J. Gubicza, 'On the strengthening behavior of ultrafine-grained nickel processed from nanopowders', *Materials Science and Engineering*, A 527 (2010) 3227–3235.
63. Q. Wei, T. Jiao, K.T. Ramesh, E. Ma, L.J. Kecskes, L. Magness, R. Dowding, V.U. Kazykhanov and R.Z. Valiev, 'Mechanical behavior and dynamic failure of high-strength ultrafine grained tungsten under uniaxial compression', *Acta Materialia*, 54 (2006) 77–87.
64. E. Ma, 'Eight routes to improve the tensile ductility of bulk nanostructured metals and alloys', *Journal of the Minerals, Metals and Materials Society*, 58 (2006) 49–53.

65. B. Ahn, E.J. Lavernia and S.R. Nutt, 'Dynamic observations of deformation in an ultrafine-grained Al–Mg alloy with bimodal grain structure', *Journal of Materials Science* 43 (2008) 7403–7408.
66. Y. Wang, M. Chen, F. Zhou and E. Ma, 'High tensile ductility in a nanostructured metal', *Nature*, 419 (2002) 912–915.
67. Y.H. Zhao, T. Topping, J.F. Bingert, J.J. Thornton, A.M. Dangelewicz, Y. Li, W. Liu, Y.T. Zhu, Y.Z. Zhou and E.J. Lavernia, 'High tensile ductility and strength in bulk nanostructured nickel', *Advanced Materials*, 20 (2008) 3028–3033.
68. L. Lu, Y. Shen, X. Chen, L. Qian and K. Lu, 'Ultrahigh strength and high electrical conductivity in copper', *Science*, 304 (2004) 422–426.
69. R.Z. Valiev, I.V. Alexandrov, Y.T. Zhu and T.C. Lowe, 'Paradox of Strength and Ductility in Metals Processed by Severe Plastic Deformation', *Journal of Materials Research*, 17 (2002) 5–8.
70. S.V. Dobatkin, J.A. Szpunar, A.P. Zhilyaev, J.-Y. Cho and A.A. Kuznetsov, 'Effect of the route and strain of equal-channel angular pressing on structure and properties of oxygen-free copper', *Materials Science and Engineering*, A 462 (2007) 132–138.
71. F. Dalla Torre, R. Lapovok, J. Sandlin, P.F. Thomson, C.H.J. Davies and E.V. Pereloma 'Microstructures and properties of copper processed by equal channel angular extrusion for 1–16 passes', *Acta Materialia*, 52 (2004) 4819–4832.
72. J. Wang, Z. Horita, M. Furukawa, M. Nemoto, N.K. Tsenev, R.Z. Valiev, Y. Ma and T.G. Langdon, 'An investigation of ductility and microstructural evolution in an Al-3% Mg alloy with submicron grain size', *Journal of Materials Research*, 8 (1993) 2810–2818.

73. J. Wang, Y. Iwahashi, Z. Horita, M. Furukawa, M. Nemoto, R.Z. Valiev and T.G. Langdon, 'An investigation of microstructural stability in an Al–Mg alloy with submicrometer grain size', *Acta Materialia*, 44 (1996) 2873–2982.
74. J. Gubicza, H-Q. Bui, F. Fellah and G.F. Dirras, 'Microstructure and mechanical behavior of ultrafine-grained Ni processed by different powder metallurgy methods', *Journal of Materials Research*, 24 (2009) 217–226.
75. A.P. Zhilyaev, B.-K. Kim, J.A. Szpunar, M.D. Baro and T.G. Langdon, 'The microstructural characteristics of ultrafine-grained nickel', *Materials Science and Engineering*, A 391 (2005) 377–389.
76. F.J. Humphreys and J.W. Martin, 'The effect of dispersed phases upon the annealing behaviour of plastically deformed copper crystals', *Philosophical Magazine*, 16 (1968) 927–957.
77. L.J. Kecskes, K.C. Cho, R.J. Dowding, B.E. Schuster, R.Z. Valiev and Q. Wei, 'Grain size engineering of bcc refractory metals: Top-down and bottom-up—Application to tungsten', *Materials Science and Engineering*, A 467 (2007) 33–43.
78. G. Dirras, J. Gubicza, S. Ramtani, Q.H. Bui and T. Szilágyi, 'Microstructure and mechanical characteristics of bulk polycrystalline Ni consolidated from blends of powders with different particle size', *Materials Science and Engineering*, A 527 (2010) 1206–1214.

6

Defect structure and mechanical properties of metal matrix–carbon nanotube composites

Abstract: Carbon nanotubes (CNTs) are promising disperse phase in metal matrix composites because their high aspect ratio, elastic modulus and strength may yield to an increase of both Young's modulus and strength of the metallic ultrafine-grained matrix. In this chapter, the processing methods, the defect structure and the mechanical properties of metal matrix–CNT composites are overviewed. It is revealed that the high dispersity of CNTs in the matrix is the most important criterion for processing composites with high strength and good ductility. The correlation between the flow stress and the dislocation density for Cu–CNT composites suggests that the CNT fragments strengthen the composite rather indirectly via the increase of the dislocation density.

Key words: carbon nanotube, composite, dislocations, twinning, strength, ductility.

6.1 Processing of metal matrix–carbon nanotube composites

Carbon nanotubes (CNTs) have unique structure that gives special properties, such as: (i) extremely high elastic modulus

(~1 TPa); (ii) large strength (~40 GPa); (iii) high aspect ratio (~100–1,000); (iv) very low density (1.5–2 g/cm^3); (v) excellent chemical and thermal stabilities (up to 750°C and 2,800°C in air and vacuum, respectively); and (vi) high thermal and electrical conductivities (thermal conductivity about twice as high as diamond, electric-current-carrying capacity 1,000 times higher than copper wires) [1, 2]. In particular, this combination of properties makes CNTs ideal candidates as advanced filler materials in composites. The addition of CNTs into polymers or ceramics can yield significant increase of their conductivity, resulting in conductive or antistatic polymer or ceramic composites. Even at low volume fractions of CNTs, they form a continuous network in the matrix, i.e. they have low percolation thresholds due to their high aspect ratio. For comparison, the percolation thresholds for conventionally used carbon black filler materials and CNTs are ~16 and ~0.01 vol.%, respectively. Taking advantage of their excellent conductivity and high aspect ratio, conductive polymers and ceramics can be produced with very low volume fractions of CNTs. For example, the electrical conductivity of alumina increased by a factor of 10^{15} with addition of 5.7 vol.% CNTs [2]. Another promising application of CNTs is their incorporation as reinforcing fillers into plastic [3], ceramic [4] and metal matrix [5] composites. The application of CNTs as a strengthening disperse phase in composites is very prosperous compared with other materials as they have high hardening effect even for low volume fractions due to their high aspect ratio. Although different polymer–CNT composites have been synthesised successfully, producing metal matrix–CNT composites is rather challenging mainly due to the poor interfacial bonding between CNTs and metal matrix, as well as due to the agglomeration of CNTs in the interfaces of matrix particles. In this chapter, we focus on metal matrix–CNT composites where the matrix exhibits ultrafine-grained or nanocrystalline microstructure. First, a short survey of the processing routes of metal matrix–CNT composites is given.

Metal matrix–CNT composites are usually produced by powder metallurgy techniques [6–18], although electrodeposition [19] is also applied for this purpose. The powder metallurgy way has the following steps: (i) purification of CNTs by washing with acids in order to remove the amorphous carbon particles; (ii) blending of CNTs with the powder of the metal matrix by high-energy milling; (iii) pre-compaction of the blend, e.g. by cold isostatic pressing [13, 18]; and (iv) consolidation to high density, e.g. by extrusion [10], rolling [9], high-pressure torsion (HPT) [16, 18], shock-wave consolidation [7] or sintering [15]. In some cases, the purification and pre-compaction steps are missing. For instance, pre-compaction is not needed when the blend of metallic powders and CNTs are consolidated by plasma spray forming [20, 21]. In this case, the metallic powder particles act as carrier of the lightweight CNTs during the spraying. The powder particles melt in the high-temperature plasma and are directed towards a rotating mandrel, where they deposit and rapidly cool, forming the desired shape as depicted in Figure 6.1 [20]. It has also been shown that a modification of the accumulative roll-bonding may be used to disperse CNTs into an aluminium alloy [22]. After degreasing the aluminium and wire-brushing the surfaces, the sheets were coated once with the concentrated CNT suspension and allowed to dry. The sheets were then stacked together making a 2.8-mm-thick sandwich, which was accumulatively roll-bonded four times at room temperature (RT) [22].

Figure 6.1 **Schematic of plasma spray forming of a blend of Al-Si powder and CNTs**

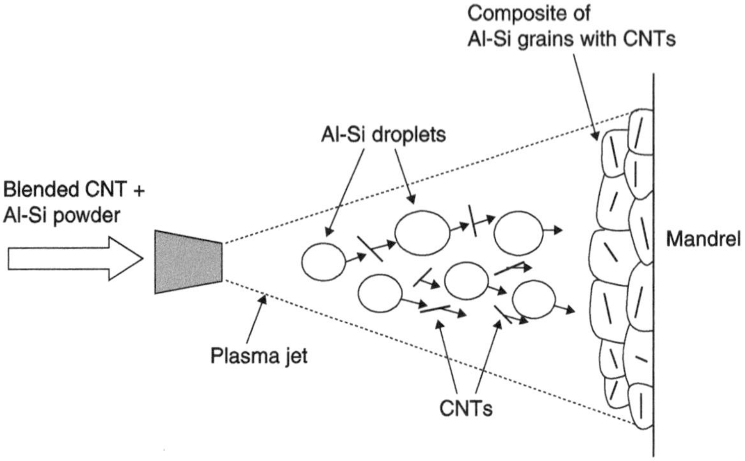

Source: After Laha et al. (2004) [20] with permission of Elsevier.

Metal-based CNT composites prepared by powder metallurgy process often show lower mechanical properties than expected, which is attributed to CNT agglomeration and also to some porosity in the composites. It is confirmed that the key issue to enhance the strength of CNT/metal nanocomposite is the homogeneous distribution of CNTs during the blending step [6]. The inhomogeneous dispersion of nanotubes in the blends yields clustering of CNTs, as shown in Figure 6.2 [21]. CNT clusters prevent sufficient bonding between the matrix particles, leading to fine pores and cracks in the nanocomposite. Subsequently, these fine cracks and pores act as nucleation sites for plastic instability, which account for the significant degradation of strain to failure of metal matrix–CNT composites [21]. The dispersion of the nanotubes can be improved by increasing the energy of milling during the mixing step of processing [11]. Another strategy for increasing the dispersion of CNTs in the matrix material is the treatment of CNTs by surfactants [10].

Figure 6.2 **SEM image of fracture surface of plasma spray formed Al–CNT nanocomposite showing intergranular fracture and cluster of CNTs**

Source: Reprinted from Laha et al. (2009) [21] with permission from Elsevier.

For instance, sodium dodecyl sulphate (SDS) surfactant decreases the van der Walls force of attraction between CNTs. When ionic SDS is used as surfactant, the negative charges in SDS micelles that absorb on nanotubes prevent re-aggregation of CNTs, thereby dispersing the CNTs in the matrix [10].

Functionalisation of nanotubes by the atoms of the metal matrix is a very effective way to reach the high dispersity of CNTs in the matrix [15]. Additionally, in the functionalised nanotubes there is a strong interfacial bond between the atoms of CNTs and the matrix that may lead to an enhancement in mechanical properties. This 'molecular level mixing' procedure involves the following steps: (i) attaching functional groups (e.g. $-COOH$, $-COCl$ or $-NO_2$ atomic group) onto the CNT surfaces; (ii) suspending functionalised CNTs in a solvent (e.g. in water or ethanol); (iii) mixing a salt containing Cu ions (e.g. copper acetate: $Cu(CH_3COO)_2$) with the CNT suspension, which results in reaction between the Cu ions and the functional groups (see Figure 6.3); (iv) drying the solution by heating at 100–250°C in air in order to remove the solvent and the ligands and oxidise Cu on CNTs to form copper-oxide powder; and (v) calcination and reduction to Cu–CNT composite in hydrogen atmosphere [15]. This process produces Cu–CNT composite powders, whereby the CNTs are homogeneously implanted within the Cu powders, as depicted schematically in Figure 6.4. Finally, the Cu–CNT composite powders are consolidated by spark plasma sintering. As this process mixes Cu and CNTs at the molecular level, the dispersity of CNTs in the sintered composite is very high. The CNTs form a homogeneously distributed network within the Cu grains [15].

Figure 6.3 **Schematic figure depicting the attachment of Cu ions to the functional groups on the surface of a CNT for the 'molecular level mixing' process of Cu–CNT composite powders**

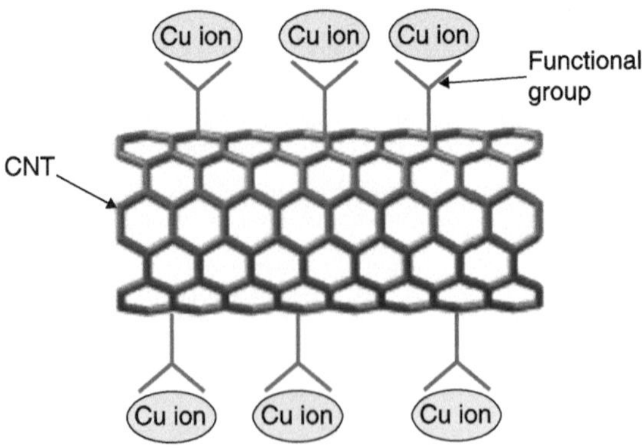

Figure 6.4 Schematic depiction of the Cu–CNT composite powder processed by 'molecular level mixing'

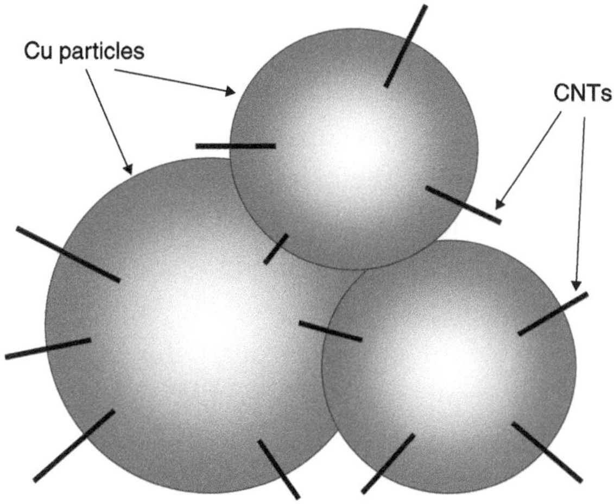

6.2 Morphology of CNTs and porosity in nanotube composites

CNTs are used as a disperse phase in metal matrix nanocomposites with the fraction of 0.5–12 vol.%. In the x-ray diffractograms obtained on metal–CNT composites processed by powder metallurgy, peaks of CNTs are often not detected. This can be explained by the very small amount of CNTs and their fragmentation during blending with the metallic powders by high-energy milling and in further consolidation by severe plastic deformation. As an example, Figure 6.5 shows a high-resolution transmission electron microscopy (HRTEM) image obtained on a Cu–CNT composite processed from micron-sized Cu particles and 3 vol.% multi-walled CNTs (MWNTs) by high-energy milling and subsequent two-step consolidation procedure including cold isostatic pressing and HPT at RT [18].

Complete MWNTs were not observed in the consolidated sample. At the same time, small graphite-like fragments were revealed from the interlayer spacing (about 0.34 nm). Both straight and curved CNT fragments were observed as indicated by the arrows in Figure 6.5. HRTEM images of other Cu–CNT composites consolidated by HPT indicate the presence of CNTs or their fragments both at the grain boundaries and inside the Cu grains [17].

It was found for Al–CNT composites that during the milling step, the impact energy of the balls could induce the infiltration of Al atoms through the inherent atomic defects on the wall of MWNTs [9]. After milling for 6 hours, the nanotubes seem to be filled with Al atoms, leading to the irregular shape of MWNTs [9]. As a

Figure 6.5 HRTEM image taken at the half-radius of the Cu–CNT-RT disk. The arrows indicate fragments of CNTs

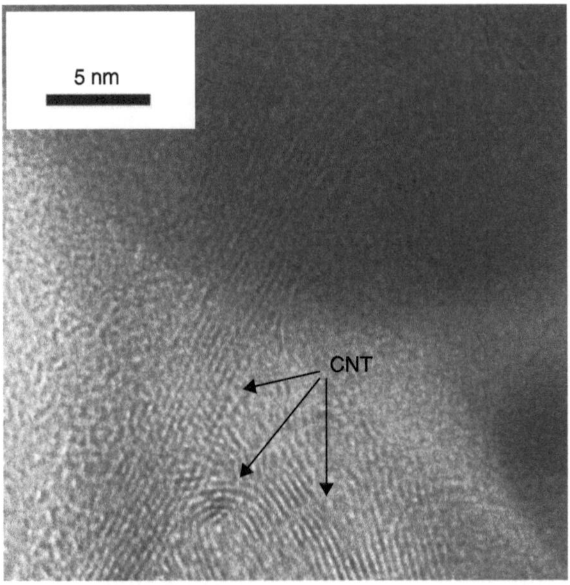

Source: Reprinted from Jenei et al. (2011) [18] with permission from Elsevier.

consequence, the diameter of MWNTs increased from ~10 nm to ~35 nm.

In some cases, besides the metal matrix and CNTs, other phases are also detected in the consolidated composite. For instance, in Cu–CNT composites processed by HPT x-ray diffraction analysis revealed the presence of copper-oxide [18]. In the initial Cu powder, additional phase besides Cu was not detected. At the same time, in the centre of the HPT disks processed from the blend of Cu powder and CNTs at both RT and 373 K a small amount of Cu_2O with a relative intensity of 1% was observed. At the half radius and the periphery of these disks oxide phase was not detected [18]. Most probably, the Cu_2O phase was formed by surface oxidation of the Cu particles during the high-energy mixing process of the Cu powder and CNTs. This oxide phase may be decayed due to severe plastic deformation by HPT, which can explain its absence at the half-radius and the periphery of the disks where the imposed strain is much higher than in the centre. In Al–CNT composites, Al_4C_3 carbides are formed in the reaction of CNTs and Al slightly above the melting temperature of Al [13]. Interfacial Al_4C_3 compound deteriorates the mechanical properties of Al–CNT composites, therefore the reaction of Al and CNTs should be avoided and the fabrication temperature of Al–CNT composites should be kept below the Al liquidus [13].

Most of the processes of metal–CNT composites are based on powder metallurgy, therefore the remaining porosity is an important factor in the determination of the mechanical performance of composites. The porosity of the consolidated samples is usually determined by density measurements using Archimedes' principle. According to the experimental results, the remaining porosity in consolidated metal matrix–CNT composites is usually between 1 and 3 vol.% [e.g. 18].

6.3 Defect structure of metal–CNT composites

The effect of CNTs on the lattice defect structure in metal matrix composites is illustrated on the example of Cu–CNT composites consolidated by HPT at RT and 373 K [18]. 99.5% purity Cu powder with particle sizes less than 44 μm and 3 vol.% MWCNTs were mixed by high-energy milling. The diameter and the length of CNTs produced by catalytic chemical vapour deposition (CCVD) were 5–20 nm and 1–10 μm, respectively. The powder blend was pre-compacted by cold isostatic pressing. The pre-compacted disks were consolidated by HPT at RT and 373 K. The applied pressure and the number of revolutions were 2.5 GPa and 10, respectively, for all the samples [18]. In order to study the effect of CNTs on microstructure, an additional sample was consolidated solely from the Cu powder at RT by the same way as in the case of the composite specimen. In the following, the samples processed from pure Cu, the blend of Cu and CNTs at RT and 373 K are denoted as Cu, Cu–CNT-RT and Cu–CNT-373, respectively. The microstructure of the HPT-processed samples was investigated at the centre, half-radius and periphery by x-ray line profile analysis. The area-weighted mean crystallite size ($<x>_{area}$), the dislocation density (ρ) and the twin boundary frequency (β) were plotted in Figures 6.6a, 6.6b and 6.6c, respectively. Usually, the crystallite size is larger while the dislocation density and the twin boundary frequency are smaller in the centre than at the half-radius or periphery due to the smaller imposed torsional strain. The curves lying on the measured data points in Figures 6.6a–6.6c are only visual guides. They reach the saturation values of the microstructural parameters between the centre and the half-radius as suggested by the hardness investigations (see section 6.4 in this chapter). For comparing

Figure 6.6 **The area-weighted mean crystallite size, $\langle x \rangle_{area}$ (a), the dislocation density, ρ (b) and the twin boundary frequency, β (c) at the centre, half-radius and periphery of the HPT-processed Cu, Cu–CNT-RT and Cu–CNT-373 disks**

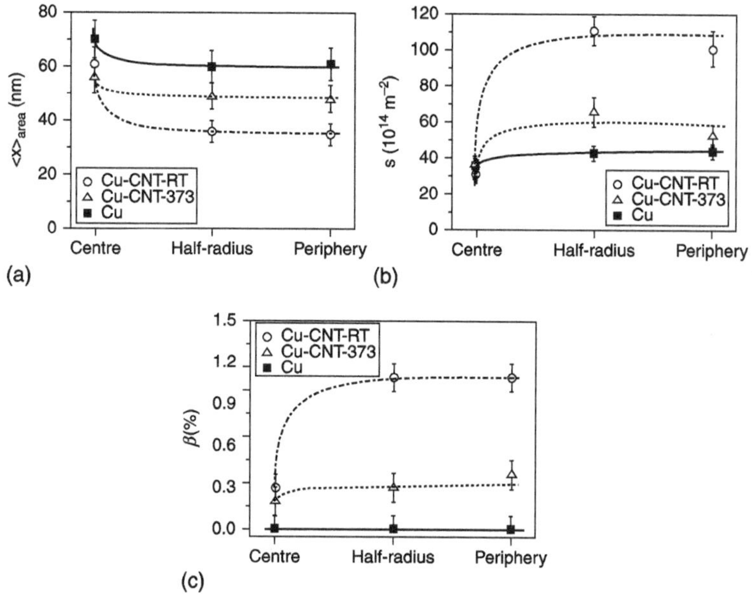

Source: Reprinted from Jenei et al. (2011) [18] with permission from Elsevier.

the three samples, it is reasonable to use the values of the microstructural parameters determined at the half-radius of the disks.

The crystallite size, the dislocation density and the twin boundary frequency measured at the half-radius in sample Cu are 60 ± 6 nm, 43 ± 4 × 10^{14} m^{-2} and 0.0 ± 0.1%, respectively [18]. In sample Cu the mean grain size is 173 nm as determined from dark-field TEM images (see Figure 6.7a). It is noted that in the case of sample Cu, the parameters of the microstructure only slightly change along the radius of the disk, indicating that the microstructure achieved the near saturation state in all locations of the sample. It is also worth noting that the crystallite size determined by x-ray line profile analysis is smaller than the grain size observed by TEM. This phenomenon has usually been observed for plastically deformed metals [23] and it can be attributed to the fact that the crystallite size determined from x-ray line profiles corresponds essentially to the mean size of cells/subgrains, which is usually smaller than the conventional grain size measured in metals by electron microscopy methods [24, 25].

Figure 6.7 Dark-field TEM images for samples (a) Cu, (b) Cu–CNT-RT and (c) Cu–CNT-373. Some twin boundaries in sample Cu–CNT-RT are indicated by white arrows in (d)

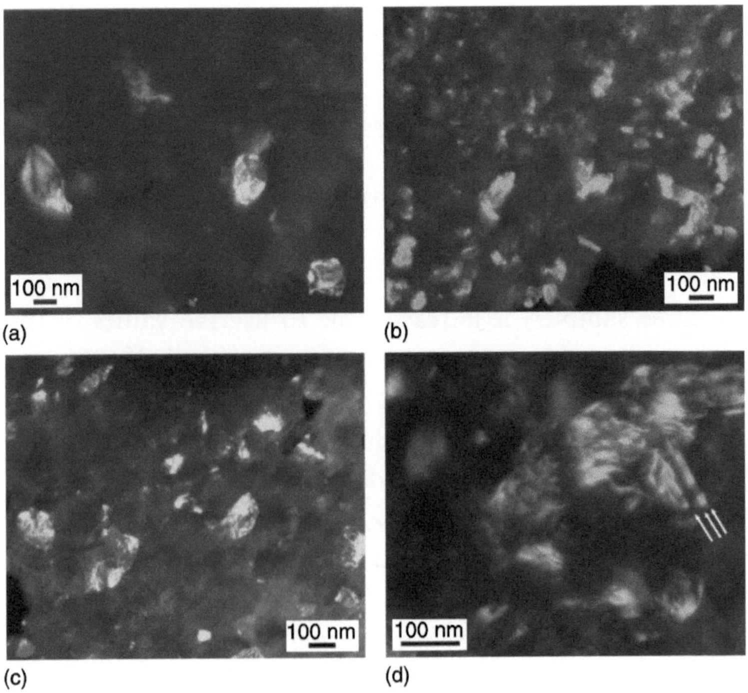

Source: Reprinted from Jenei et al. (2011) [18] with permission from Elsevier.

The mean crystallite size of the composite sample Cu–CNT-RT (36 ± 4 nm) is half as large as that of pure Cu (60 ± 6 nm), while the dislocation density (111 ± 10 × 10^{14} m^{-2}) is three times higher in the former sample at the half-radius of the disks (see Figures 6.6a and 6.6b). Figure 6.7b shows a dark-field TEM image of the grain structure for sample Cu–CNT-RT [18]. The average grain size in this specimen is 74 nm, which is about half of the value obtained for pure Cu. The smaller crystallite and grain sizes as well as the much higher dislocation density in sample Cu–CNT-RT can be explained by the pinning effect of CNT fragments on lattice defects (dislocations and grain boundaries) during HPT, which is especially significant where the imposed strain is higher, i.e. at the half-radius and the periphery of the disk. The principle of producing nanostructures by HPT is to form subgrain boundaries via dislocation accumulation and rearrangements under severe shear strain. These subgrain boundaries then develop into high-angle grain boundaries, resulting in grain refinement. At the same time, dynamic recovery and recrystallisation reduce the dislocation density and may increase the grain size. The balance of these two opposite processes determines the final grain size. With the incorporated CNTs, the dislocation motion is blocked at CNT–Cu interfaces, therefore the dislocation accumulation is enhanced. As a consequence, the addition of CNTs leads to a decrease in grain size [17]. Figure 6.6c shows that in contrast with the lack of observable twin boundaries in pure Cu, in sample Cu–CNT-RT the twin boundary frequency is high (1.1 ± 0.1%). The pinning effect of CNT fragments in sample Cu–CNT-RT hinders the escape of dislocations from pile-ups during HPT, resulting in high stresses at glide obstacles, such as Lomer-Cottrel barriers and grain boundaries. If the local stresses at these obstacles exceed the critical stress for twin nucleation, deformation twins are formed (see Chapter 3). Most probably, the small grain size has also contributed to the evolution of the very high twin boundary frequency. As was shown in previous papers [e.g. 26], the reduction in grain size of Cu led to an increasing contribution of twinning to plasticity at the expense of dislocation activity (see also Chapter 5). Some twin boundaries in sample Cu–CNT-RT are shown in Figure 6.7d.

Figures 6.6a–6.6c reveal that when the Cu–CNT composite was processed by HPT at 373 K the crystallite size (49 ± 5 nm) is larger by 36%, while the dislocation density ($66 \pm 8 \times 10^{14}$ m^{-2}) and the twin boundary frequency (0.3 ± 0.1%) are one-half and one-third of the values obtained after processing at RT, respectively. This can be explained by the higher mobility of lattice defects at 373 K, yielding their easier annihilation during HPT-straining. Figure 6.7c shows a dark-field TEM image of the microstructure in sample Cu–CNT-373. The average grain size is 83 nm, which is larger than the value determined for sample Cu–CNT-RT [18].

6.4 Correlation between defect structure and mechanical properties

The extremely high elastic modulus (1 TPa) of CNTs along their axes yields a very large increment of Young's modulus with their addition to the metal matrix. As an example, Figure 6.8 shows the measured Young's modulus of Al–CNT composites as a function of volume fraction of CNTs [9]. The modulus increases linearly with increasing the volume fraction of MWNTs. The experimental results are compared with the calculated Young's modulus values of the composites that are obtained from the expression [9]:

$$E_c = K_{CNT} E_{CNT} V_{CNT} + E_m (1 - V_{CNT} - V_p), \qquad (6.1)$$

Figure 6.8 **The Young's modulus and the yield strength as a function of the volume fraction of MWNTs in Al–CNT composites**

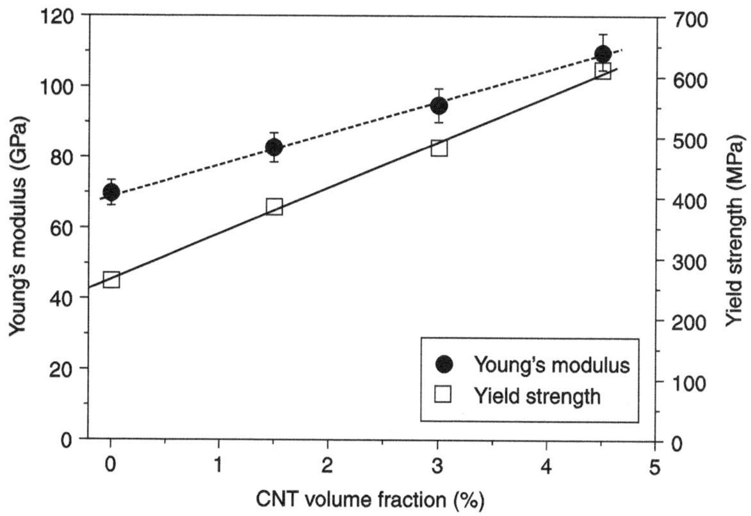

Source: The data are taken from Choi et al. (2009) [9].

where E_c, E_{CNT} and E_m are the Young's moduli of the composite, CNTs and the matrix, and V_{CNT} and V_p are the volume fractions of CNTs and pores, respectively, and K_{CNT} is a reinforcement factor. For example, if CNTs are randomly oriented within the matrix or they are aligned along the loading direction, K_{CNT} is about 0.2 or 1, respectively [9]. The dashed line in Figure 6.8 illustrates the moduli calculated from Equation 6.1 for the case of K_{CNT} = 1 taking E_{CNT} = 1 TPa and V_p = 0. The experimental and calculated values are very well matched, which can be explained by the alignment of CNTs along the loading direction. These Al–CNT composites are consolidated from powder blends by hot-rolling [9]. Most MWNTs in the final composites are aligned along the rolling direction, following the macroscopic variation of the powder container during rolling [9]. As in the tensile test the samples were loaded in the rolling direction, therefore CNTs were lying parallel to the loading direction. The Young's modulus values for different metal matrix–CNT composites are listed in Table 6.1. For comparison, the moduli of the matrix materials without CNTs are also given in the table. The ratio of the moduli of the CNT composites and their matrices are plotted as a function of volume fraction of CNTs in Figure 6.9. The higher CNT content yields larger Young's modulus irrespective of the matrix composition and the characteristics of CNTs (single- or multi-walled, diameter and length).

The pinning effect of CNTs on the lattice defects (dislocations, grain-boundaries) causes an increment in yield strength as shown in Figure 6.8 for Al–CNT composites [9]. Table 6.1 lists the yield strength obtained on different metal–CNT composites together with the values determined on the matrix materials. Figure 6.9 shows that the higher amount of CNTs usually results in larger increment in yield strength. The correlation between the yield strength and lattice defect structure was examined in detail on Cu–CNT composites consolidated by HPT [18]. The processing and the defect structure in these materials are presented in section 6.3 of this chapter. The samples processed from pure Cu, the blend of Cu and 3 vol.% CNTs at RT and 373 K are denoted as Cu, Cu–CNT-RT and Cu–CNT-373, respectively. The microhardness as a function of the distance from the centre of the HPT-processed disks is plotted in Figure 6.10a. It can be revealed that the hardness increases with increasing the distance from the centre and gets saturated already at about 20% of the radius for all the three samples. The smaller hardness in the centre can be explained by the smaller strain imposed during HPT compared with the other locations along the radius of the disk. Despite the linear dependence of the strain on the distance from the centre, the hardness varies in a non-linear manner due to decreasing work hardening

Table 6.1 The features of the matrix and the applied CNTs, the processing method, the Young's modulus, the yield and tensile strengths, and the maximum elongation for different metal matrix–CNT composites. For the Young's modulus, the yield and tensile strengths and the elongation, the values obtained without CNTs are also given in parentheses. The CNT volume fractions are taken from the original papers or calculated from the weight percentages using the densities of CNTs and the matrices

Matrix *Material/grain size*	CNT characteristics *Diameter/length/ volume fraction*	Processing method	Young's modulus [GPa]	Yield strength [MPa]	Tensile strength [MPa]	Max. elongation [%]	Reference
Al/600 nm	2–40 nm/1 μm/2 vol.%	Shock-wave consolidation	–	130 (133)	–	1.4 (6.5)	[7]
Al/100 nm	1–2 nm/–/7.3 vol.% (5 wt.%)	Ultrasonic mixing + High-pressure torsion	–	150 (100)	210 (150)	0.5 (0.8)	[8]
Al/150 nm	10 nm/5 μm/1.5 vol.%	Ball milling + Hot rolling	83 (70)	386 (262)	391 (282)	7 (13)	[9]
Al/150 nm	10 nm/5 μm/3 vol.%	Ball milling + Hot rolling	95 (70)	483 (262)	491 (282)	4 (13)	[9]
Al/150 nm	10 nm/5 μm/4.5 vol.%	Ball milling + Hot rolling	110 (70)	610 (262)	615 (282)	2 (13)	[9]

(Continued)

Table 6.1 **The features of the matrix and the applied CNTs, the processing method, the Young's modulus, the yield and tensile strengths, and the maximum elongation for different metal matrix–CNT composites (*Continued*)**

Matrix *Material/grain size*	CNT characteristics *Diameter/length/volume fraction*	Processing method	Young's modulus [GPa]	Yield strength [MPa]	Tensile strength [MPa]	Max. elongation [%]	Reference
Al/1 μm	10 nm/5–15 μm/0.75 vol.% (0.5 wt.%)	Mechanically mixed/Cold uniaxial pressing/Free-sintering/Hot extrusion	62	114 (91)	122 (98)	2.2 (2.5)	[10]
Al/1 μm	10 nm/5–15 μm/1.5 vol.% (1 wt.%)	Mechanically mixed/Cold uniaxial pressing/Free-sintering/Hot extrusion	66	139 (91)	151 (98)	3.0 (2.5)	[10]
Al/1 μm	10 nm/5–15 μm/3 vol.% (2 wt.%)	Mechanically mixed/Cold uniaxial pressing/Free-sintering/Hot extrusion	75	176 (91)	184 (98)	2.7 (2.5)	[10]

Al/–	140 nm/3–4 μm/0.75 vol.% (0.5 wt.%)	Blending by shaker mixer and planetary mill + Hot rolling	60 (50)	100 (70)	144 (130)	19 (25)	[11]
Al/–	140 nm/3–4 μm/1.5 vol.% (1 wt.%)	Blending by shaker mixer and planetary mill + Hot rolling	40 (50)	68 (70)	105 (130)	7.5 (25)	[11]
Al/–	140 nm/3–4 μm/3 vol.% (2 wt.%)	Blending by shaker mixer and planetary mill + Hot rolling	47 (50)	44 (70)	62 (130)	2 (25)	[11]
Al/–	40–100 nm/10 μm/0.75 vol.% (0.5 wt.%)	Milling + Pressureless sintering + Hot extrusion	–	140 (160)	180 (105)	12 (20)	[12]
Al/–	40–100 nm/10 μm/2.2 vol.% (1.5 wt.%)	Milling + Pressureless sintering + Hot extrusion	–	190 (160)	240 (105)	14 (20)	[12]
Al/–	40–100 nm/10 μm/3 vol.% (2 wt.%)	Milling + Pressureless sintering + Hot extrusion	–	180 (160)	250 (105)	16 (20)	[12]
93.5% Al-4.4% Cu-1.5% Mg-0.6% Mn (2024Al)/200 nm	20–40 nm/10 μm/1.5 vol.% (1 wt.%)	Ball milling + Cold isostatic pressing + Extrusion	88 (71)	336 (289)	474 (384)	3 (16.5)	[13]

(Continued)

Table 6.1 **The features of the matrix and the applied CNTs, the processing method, the Young's modulus, the yield and tensile strengths, and the maximum elongation for different metal matrix–CNT composites (*Continued*)**

Matrix *Material/grain size*	CNT characteristics *Diameter/length/ volume fraction*	Processing method	Young's modulus [GPa]	Yield strength [MPa]	Tensile strength [MPa]	Max. elongation [%]	Reference
Ni/28 nm	10–30 nm/1–2 μm/–	Electrodeposition	–	–	1,475 (1,162)	2.09 (2.39)	[19]
90% Mg-9% Al-0.6% Zn-0.4% Mn (AZ91D)/10 μm	100 nm/5 μm/ 0.5 vol.%	Ball milling + Hot press + Extrusion	43 (40)	281 (232)	383 (315)	6 (14)	[14]
90% Mg-9% Al-0.6% Zn-0.4% Mn (AZ91D)/10 μm	100 nm/5 μm/ 1 vol.%	Ball milling + Hot press + Extrusion	49 (40)	295 (232)	388 (315)	5 (14)	[14]
90% Mg-9% Al-0.6% Zn-0.4% Mn (AZ91D)/10 μm	100 nm/5 μm/ 3 vol.%	Ball milling + Hot pressing + Extrusion	51 (40)	284 (232)	361 (315)	3 (14)	[14]

90% Mg-9% Al-0.6% Zn-0.4% Mn (AZ91D)/10 μm	100 nm/5 μm/ 5 vol.%	Ball milling + Hot pressing + Extrusion	51 (40)	277 (232)	307 (315)	1 (14)	[14]
Cu/250 nm	40 nm/1–2 μm/ 5 vol.%	Ball milling + Spark Plasma Sintering + Cold rolling	127 (70)	149 (135)	220 (175)	8 (14)	[6]
Cu/250 nm	40 nm/1–2 μm/ 10 vol.%	Ball milling + Spark Plasma Sintering + Cold rolling	137 (70)	197 (135)	281 (175)	6.5 (14)	[6]
Cu/100 nm	40 nm/2 μm/ 5 vol.% (1 wt.%)	CNT surface functionalisation by Cu + Spark plasma sintering	114 (82)	350 (110)	–	–	[15]
Cu/100 nm	40 nm/2 μm/10 vol.% (2.2 wt.%)	CNT surface functionalisation by Cu + Spark plasma sintering	135 (82)	450 (110)	–	–	[15]
Cu/22 nm	10 nm/1 μm/4.76 vol.% (1 wt.%)	Ball milling + High-pressure torsion	117 (91)	1,125 (738)	–	–	[16,17]
Cu/74 nm	5–20 nm/1–10 μm/3 vol.%	High-energy milling + Cold Isostatic pressing + High-pressure torsion	–	770 (580)	–	–	[18]

Figure 6.9 The ratio of the Young's moduli, the yield and tensile strengths and the maximum elongations in tension obtained for CNT-composites and their pure matrices. The data are taken from Table 6.1

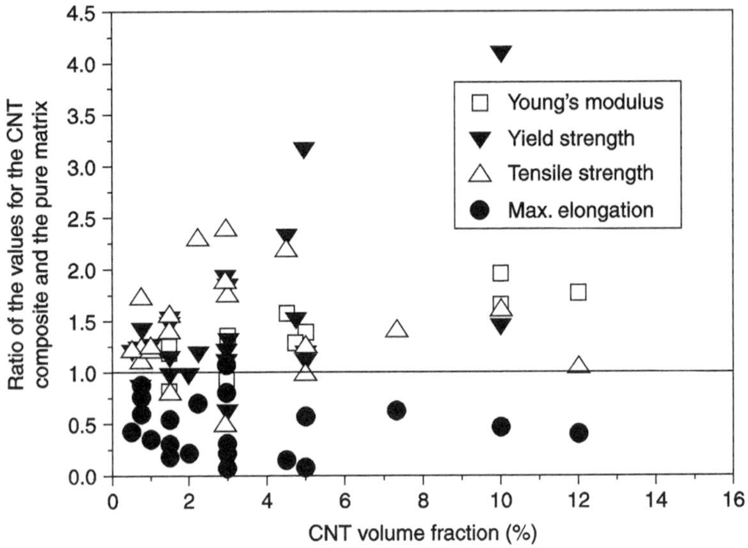

with strain, which is in agreement with previous observations [8, 16]. It is noted that although the saturation hardness values of the Cu–CNT composites are higher than for the pure Cu sample, close to the disk centre this relation is reversed (see Figure 6.10a). In section 6.3, it was shown that in the three samples near the centre the dislocation density values are close to each other, therefore they have similar hardening effects. At the same time, the significant remaining porosity in the composite samples (about 3%) results in hardness reduction. Far from the centre, the hardening effect of the much larger dislocation density in the composites overwhelms the influence of porosity, leading to higher hardness than for the pure Cu sample. The hardness measured at 350 K also saturated already at about 20% of the radius for all the three samples (not shown here). The saturation hardness values determined at the half-radius of the disks for RT and 350 K are plotted in Figure 6.10b. For all the three specimens, the hardness reduced by about 20% when the temperature increased from RT to 350 K. This means that the relative hardening effect of CNTs was maintained at least until 350 K. The smaller grain size as well as the larger dislocation and twin boundary densities may explain the much higher hardness of specimen Cu–CNT-RT compared with pure Cu. The smaller defect densities and larger grain size in sample Cu–CNT-373 may yield smaller hardness compared with specimen Cu–CNT-RT.

The parameters of the microstructure and the yield strength are usually correlated using the Taylor and/or Hall-Petch formulas. The former and the latter equations express the hardening effects caused by dislocations and boundaries, respectively. The yield strength values of the samples Cu, Cu–CNT-RT and Cu–CNT-373 were determined at the centre, half-radius and periphery as one-third of the hardness obtained at RT. These values are compared with the yield strength calculated from the dislocation density using the Taylor equation and corrected for the porosity as [27]:

$$\sigma = (\sigma_0 + \alpha M^T G b \rho^{1/2}) \exp(-0.05 \cdot V_p) \qquad (6.2)$$

where σ_0 is the friction stress (35 MPa [28]), α is a constant (0.22 [29]), G is the shear modulus (47 GPa [30]), b is the length of the Burgers vector (0.256 nm), M^T is the Taylor factor (3.06 was selected as strong texture was not observed in the samples) and V_p is the porosity in percentage terms (0% for Cu and 3% for Cu–CNT-RT and Cu–CNT-373). The measured and the calculated yield strength values are in good agreement, as shown in Figure 6.11. The yield strength was also calculated at the half-radius of the disks from the

Figure 6.10 The microhardness (HV) as a function of the distance from the centre of the HPT-processed Cu, Cu–CNT-RT and Cu–CNT-373 disks (a). The hardness values determined at the half-radius of the disks at RT and 350K (b)

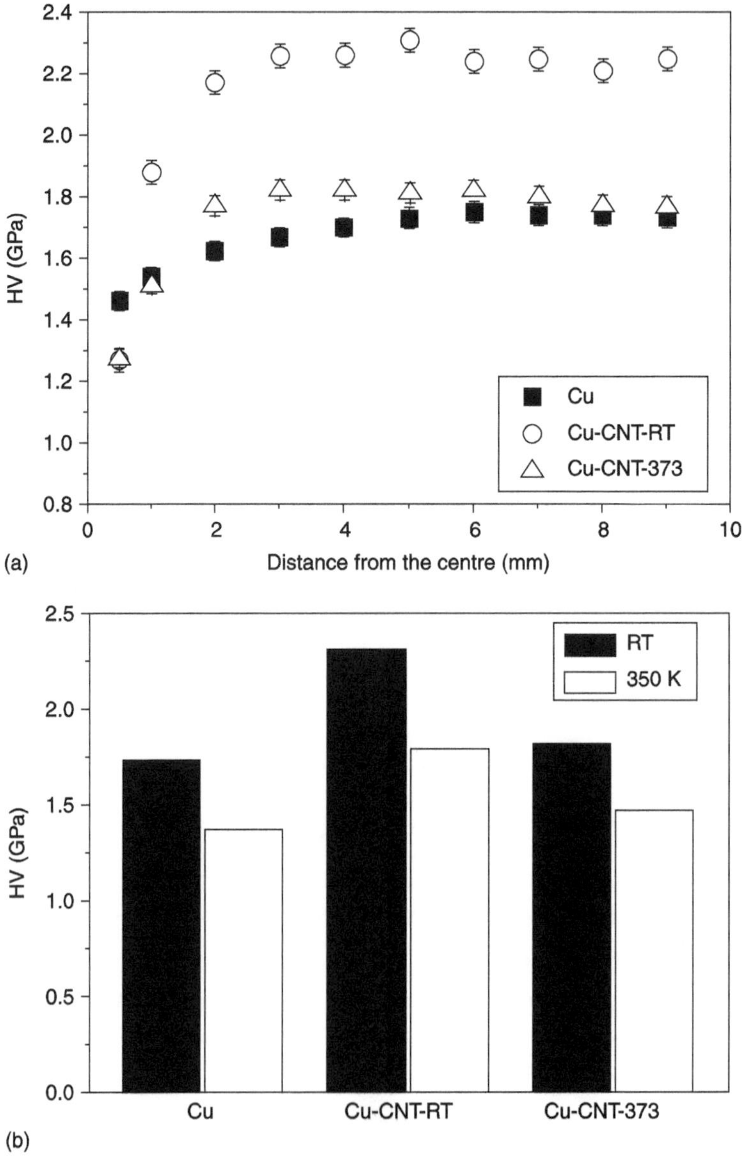

Source: Reprinted from Jenei et al. (2011) [18] with permission from Elsevier.

Figure 6.11 The calculated yield strength versus the measured values obtained as one-third of the hardness at the centre, half-radius and periphery of the HPT-processed Cu, Cu–CNT-RT and Cu–CNT-373 disks. The yield strength was calculated from the grain size obtained by TEM using the Hall-Petch formula and also from the dislocation density measured by x-ray line profile analysis using the Taylor equation

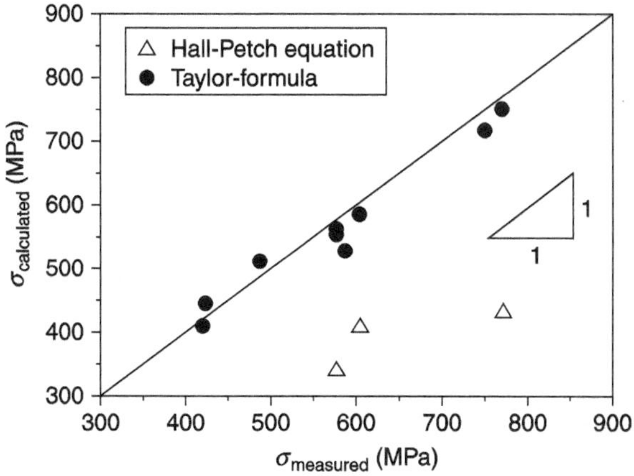

Source: Reprinted from Jenei et al. (2011) [18] with permission from Elsevier.

grain size (d) determined by TEM using the Hall-Petch formula corrected for the porosity:

$$\sigma = (\sigma_0 + kd^{-1/2}) \exp(-0.05 \cdot V_p), \tag{6.3}$$

where k is the Hall-Petch slope (~4,100 MPa·nm$^{1/2}$ for UFG Cu [31]). Figure 6.11 demonstrates that the yield strength determined from the grain size is much less than the experimental values, in agreement with previous studies on Cu–CNT composites [16].

It seems that although the Taylor equation takes only the interaction between dislocations into account, it is capable solely to estimate the yield strength of the microstructure in the HPT-consolidated Cu and Cu–CNT composites. This can be explained by the severely deformed microstructures in these samples containing extremely high dislocation densities (in the order of 10^{16} m^{-2}). First, in severely deformed structures the majority of grain and subgrain boundaries are built up from dislocations, therefore their strengthening contributions can be described by a Taylor-type equation, as pointed out theoretically by Hughes and Hansen [32]. Secondly, the stress needed for passing a dislocation through the high density dislocation structure is so large that it is also enough for crossing the twin boundaries, therefore they have no additional strengthening contribution. It is also interesting to note that the CNT fragments would be expected to have an additional dispersion strengthening contribution to the flow stress, but this effect cannot be detected since the yield strength determined by the Taylor formula agrees well with the measured values. This observation can be explained by the severe plastic deformation during sample production as follows. In HPT-processing, a very large number of dislocations passed through the CNT fragments, resulting in Orowan loops around them. When the consolidated samples are deformed subsequently in hardness testing, the gliding dislocations interact directly with these loops instead of the CNT fragments, therefore this strengthening effect is included in the Taylor equation. Therefore, the CNT fragments in Cu have no direct strengthening effect but rather they harden via the increase of the dislocation density.

Although the incorporation of CNTs into metallic matrices increases the elastic modulus and the yield strength with the nanotube volume fraction, the elongation to failure during tension is usually reduced with the addition of CNTs into the matrix materials. Table 6.1 shows the maximum elongation in tension for different metal matrix–CNT composites. The incomplete bonding between CNTs and the matrix grains, as well as porosity caused by clustering of nanotubes during blending, may yield lower fracture toughness and therefore smaller ductility than for the materials without CNTs. As a consequence, the maximum elongation is very sensitive to the processing conditions and the amount of CNTs alone does not determine the ductility, as seen in Figure 6.9. A similar argument can be applied for the tensile strength (see Figure 6.9). It is noted that for a given processing technique and matrix material, the tendency of agglomeration of CNTs increases with increasing nanotube volume fraction, which yields a lower ductility for higher CNT contents (see Table 6.1). The high dispersity of CNTs in the matrix can be achieved by 'molecular level mixing' of nanotubes and the matrix material [10] or by decreasing the adhesion between CNTs using surfactants [15], as discussed in section 6.1 of this chapter. The application of these processing steps increases the tensile strength and the strain to failure of the composites.

6.5 References

1. E.T. Thostensona, Z. Renb and T.-W. Chou, 'Advances in the science and technology of carbon nanotubes and their composites: a review', *Composites Science and Technology*, 61 (2001) 1899–1912.
2. V. Viswanathan, T. Laha, K. Balani, A. Agarwal and S. Seal, 'Challenges and advances in nanocomposite processing techniques', *Materials Science and Engineering*, R 54 (2006) 121–285.
3. J.N. Coleman, U. Khan, W.J. Blau and Y.K. Gun'ko, 'Small but strong: A review of the mechanical properties of carbon nanotube polymer composites', *Carbon*, 44 (2006) 1624–1652.

4. Cs. Balazsi, F. Konya, L. Weber, L. Biro and P. Arato, 'Preparation and characterization of carbon nanotube reinforced silicon nitride composites', *Materials Science and Engineering*, C 23 (2003) 1133–1137.
5. H. Li, A. Misra, Z. Horita, C.C. Koch, N.A. Mara, P.O. Dickerson and Y. Zhu, 'Strong ductile nanostructured Cu-carbon nanotube composite', *Applied Physics Letters*, 95 (2009) 071907.
6. K.T. Kim, S.I. Cha, S.H. Hong and S.H. Hong, 'Microstructures and tensile behaviour of carbon nanotube reinforced Cu matrix nanocomposites', *Materials Science and Engineering*, A 430 (2006) 27–33.
7. W. Salas, N.G. Alba-Baena and L.E. Murr, 'Explosive Shock-Wave Consolidation of Aluminium Powder/ Carbon Nanotube Aggregate Mixtures: Optical and Electron Metallography', *Metallurgical and Materials Transactions*, A 38 (2007) 2928–2935.
8. T. Tokunaga, K. Kaneko and Z. Horita, Production of aluminium-matrix carbon nanotube composite using high pressure torsion. *Materials Science and Engineering*, A 490 (2008) 300–304.
9. H. Choi, J. Shin, B. Min, J. Park and D. Bae, 'Reinforcing effects of carbon nanotubes in structural aluminium matrix nanocomposites', *Journal of Materials Research*, 24 (2009) 2610–2616.
10. I. Sridhar and K.R. Narayanan, 'Processing and characterization of MWCNT reinforced aluminium matrix composites', *Journal of Materials Science*, 44 (2009) 1750–1756.
11. A.M.K. Esawi and M.A. El Borady, 'Carbon nanotube-reinforced aluminium strips', *Composites Science and Technology*, 68 (2008) 486–492.
12. R. Pérez-Bustamante, C.D. Gómez-Esparza, I. Estrada-Guel, M. Miki-Yoshida, L. Licea-Jiménez, S.A. Pérez-

García and R. Martínez-Sánchez, 'Microstructural and mechanical characterization of AI-MWCNT composites produced by mechanical milling', *Materials Science and Engineering*, A 502 (2009) 159–163.

13. C. Deng, X.X. Zhang, D. Wang, Q. Lin and A. Li, 'Preparation and characterization of carbon nanotubes/ aluminium matrix composites', *Materials Letters*, 61 (2007) 1725–1728.
14. Y. Shimizu, S. Miki, T. Soga, I. Itoh, H. Todoroki, T. Hosono, K. Sakaki, T. Hayashi, Y.A. Kim, M. Endo, S. Morimoto and A. Koide, 'Multi-walled carbon nanotube-reinforced magnesium alloy composites', *Scripta Materialia*, 58 (2008) 267–270.
15. S.I. Cha, K.T. Kim, S.N. Arshad, C.B. Mo and S.H. Hong, 'Extraordinary Strengthening Effect of Carbon Nanotubes in Metal-Matrix Nanocomposites Processed by Molecular-Level Mixing', *Advanced Materials*, 17 (2005) 1377–1381.
16. H. Li, A. Misra, Y. Zhu, Z. Horita, C.C. Koch and T.G. Holesinger, 'Processing and characterization of nanostructured Cu-carbon nanotube composites', *Materials Science and Engineering*, A 523 (2009) 60–64.
17. H. Li, A. Misra, Z. Horita, C.C. Koch, N.A. Mara, P.O. Dickerson and Y. Zhu, 'Strong and ductile nanostructured Cu-carbon nanotube composite', *Applied Physics Letters*, 95 (2009) 071907.
18. P. Jenei, E.Y. Yoon, J. Gubicza, H.S. Kim, J.L. Lábár and T. Ungár, 'Microstructure and hardness of copper-carbon nanotube composites consolidated by High Pressure Torsion', *Materials Science and Engineering*, A 528 (2011) 4690–4695.
19. P.-Q. Dai, W.-C. Xu and Q.-Y. Huang, 'Mechanical properties and microstructure of nanocrystalline

nickel-carbon nanotube composites produced by electrodeposition', *Materials Science and Engineering*, A 483–484 (2008) 172–174.

20. T. Laha, A. Agarwal, T. McKechnie and S. Seal, 'Synthesis and characterization of plasma spray formed carbon nanotube reinforced aluminium composite', *Materials Science and Engineering*, A 381 (2004) 249–258.
21. T. Laha, Y. Chen, D. Lahiri and A. Agarwal, 'Tensile properties of carbon nanotube reinforced aluminium nanocomposite fabricated by plasma spray forming Composites: Part A', *Applied Science and Manufacturing*, 40 (2009) 589–594.
22. S. Salimi, H. Izadi and A.P. Gerlich, 'Fabrication of an aluminium-carbon nanotube metal matrix composite by accumulative roll-bonding', *Journal of Materials Science*, 46 (2011) 409–415.
23. Y.T. Zhu, J.Y. Huang, J. Gubicza, T. Ungar, Y.M. Wang, E. Ma and R.Z. Valiev, 'Nanostructures in Ti processed by severe plastic deformation', *Journal of Materials Research*, 18 (2003) 1908–1917.
24. T. Ungár, G. Tichy, J. Gubicza and R.J. Hellmig, 'Correlation between subgrains and coherently scattering domains', *Powder Diffraction*, 20 (2005) 366–375.
25. J. Gubicza and T. Ungár, 'Nanocrystalline materials by powder diffraction line profile analysis', *Zeitschrift fur Kristallographie*, 222 (2007) 567–579.
26. L. Balogh, G. Ribárik and T. Ungár, 'Stacking faults and twin boundaries in fcc crystals determined by x-ray diffraction profile analysis', *Journal of Applied Physics*, 100 (2006) 023512.
27. J. Luo and R. Stevens, 'Porosity-dependence of elastic moduli and hardness of 3Y-TZP ceramics', *Ceramics International*, 25 (1999) 281–286.

28. N.Q. Chinh, J. Gubicza and T.G. Langdon, 'Characteristics of face-centered cubic metals processed by equal-channel angular pressing', *Journal of Materials Science*, 42 (2007) 1594–1605.
29. J. Gubicza, N.Q. Chinh, J.L. Lábár, Z. Hegedűs, C. Xu and T.G. Langdon, 'Microstructure and yield strength of severely deformed silver', *Scripta Materialia*, 58 (2008) 775–778.
30. D.H. Chung and W.R. Buessem, in: F.W. Vahldiek, S.A. Mersol (eds), *Anisotropy of Single Crystal Refractory Compounds*, vol. 2, Plenum Press, New York, 1968, pp.217–245.
31. N. Hansen, 'Hall-Petch relation and boundary strengthening', *Scripta Materialia*, 51 (2004) 801–806.
32. D.A. Hughes and N. Hansen, 'Microstructure and strength of nickel at large strains', *Acta Materialia*, 48 (2000) 2985–3004.

7

Thermal stability of defect structures in nanomaterials

Abstract: The thermal stability of the defect structure in nanomaterials is of great importance from the point of view of practical applications of these materials. Thus, if the fine grains become coarsened or the high density of dislocations decreases during their service lifetime, their unique properties including the high strength will be lost. The high-temperature thermal stability is usually investigated by differential scanning calorimetry combined with the direct or indirect observations of the microstructure. At the same time, some nanomaterials (e.g. Cu and Ag) tend to recover/recrystallise even during storage at the processing temperature (e.g. at room temperature), which is called self-annealing. In this chapter, the thermal stability at both high and room temperatures is discussed. At high temperatures, the activation energy of recovery/recrystallisation in nanomaterials is close to the activation energy of grain-boundary diffusion. The onset temperature of recovery/recrystallisation and the released heat depend on the grain size and the defect density contrary to the activation energy. The low stacking fault energy promotes self-annealing as the recrystallised grains may be easily separated from the matrix by low-energy twin boundaries.

Key words: recovery, recrystallisation, dislocation density, vacancy annihilation, calorimetry.

7.1 High-temperature thermal stability of nanomaterials

The high-temperature thermal stability for various nanomaterials processed by severe plastic deformation (SPD) methods has been studied by differential scanning calorimetry (DSC). As an example, Figure 7.1 shows DSC thermograms taken at the heating rate of 10 K/min on 99.995% (4N5) purity Ag processed by 1, 4, 8 and 16 passes of equal-channel angular pressing (ECAP). The grain size, the dislocation density and the twin boundary frequency for these samples are presented in section 3.2 of Chapter 3. The exothermic peak evolved during heating corresponds to the annihilation of lattice defects (vacancies, dislocations, grain boundaries) during recovery and recrystallisation. Although the

Figure 7.1 DSC thermograms taken at the heating rate of 10K/min on 99.995% (4N5) purity Ag processed by 1, 4, 8 and 16 passes of ECAP

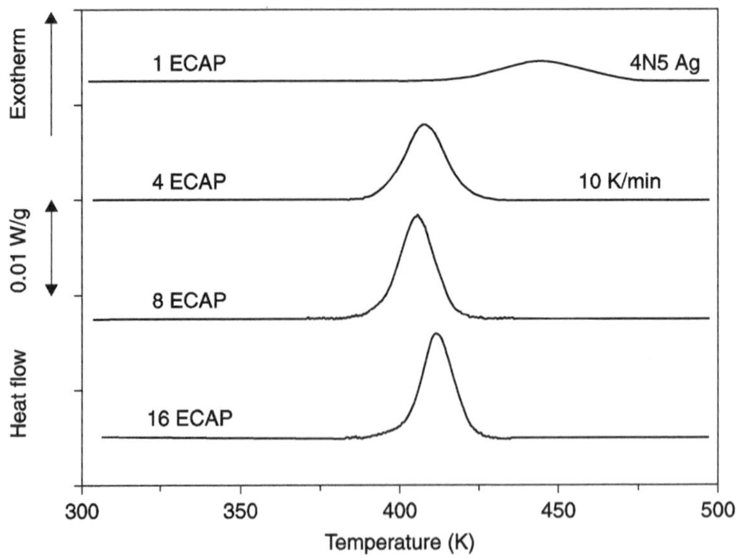

temperature of the peak maximum depends on the SPD-processing method, the imposed strain and the heating rate, its value is usually between 0.3–0.4 × T_m, where T_m is the melting point in Kelvin degrees.

The maximum of the exothermic peak in the DSC trace shifted to lower temperature values and the heat released during the annealing increased with the increase of the number of ECAP passes up to 8 passes (see Figure 7.2). These trends have also been observed for other ultrafine-grained (UFG) metals processed by different methods of SPD [1–7] and can be explained by the increase of the amount of grain boundaries due to grain refinement and the increase of densities of other lattice defects such as dislocations, stacking faults and vacancies. These defects increase the driving force for nucleation of new defect-free grains so that recrystallisation is achieved at lower temperatures. The slight decrease of the released heat and the increase of peak temperature for Ag between 8 and 16 ECAP passes are attributed to the decrease of dislocation density due to dynamic recovery processes occurring at large

Figure 7.2 **(a) The temperature of the maximum of the DSC peaks presented in Figure 7.1 for 4N5 Ag processed by different number of passes of ECAP. (b) The heat released in the DSC peaks and the activation energies determined by Kissinger method as a function of number of ECAP passes**

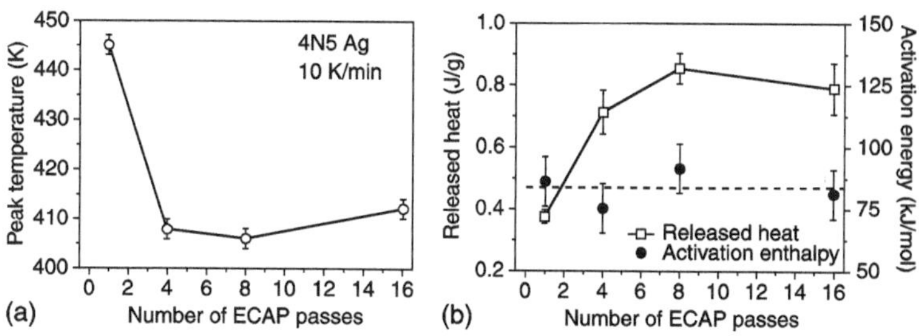

strains of ECAP. The reduction in the dislocation density at large number of ECAP passes has also been observed for other UFG metals (see also Chapter 2) [8, 9].

It was found for pure Cu and Ag samples processed by SPD that neither significant reduction in lattice defect densities nor grain-growth occurs before the exothermic peak during annealing in DSC [4]. X-ray line profile analysis investigations revealed that at the temperature corresponding to the beginning of the DSC peak, each x-ray diffraction peak profile appears to be a sum of a narrow and a broad peak, indicating a development of a bimodal grain structure [3, 4]. This microstructure consists of recovered/recrystallised grains and volumes with unchanged defect structure resulting in broad and narrow x-ray diffraction peaks, respectively. The formation of bimodal microstructure during annealing of deformed Cu has also been observed by transmission electron microscopy (TEM) [10, 11]. The variation of recovery/recrystallisation rate in different locations of nanomaterials is caused by the inhomogeneous microstructure formed during processing. It is anticipated that the annihilation of lattice defects starts at lower temperatures in the volumes having larger defect densities. In the temperature range corresponding to the exothermic peaks, both recovery (annihilation of vacancies and dislocations) and recrystallisation occur [3, 4].

The effect of the processing method on the thermal stability of UFG materials was studied on oxygen-free copper (99.98% purity) [12]. Samples were processed by 15 passes of twist extrusion (TE), 20 cycles of multi-directional forging (MDF), 25 passes of ECAP and 25 revolutions of high pressure torsion (HPT) at room temperature (RT). The details of the processing conditions can be found in Gubicza et al. (2011) [12]. The thermal stability was monitored by DSC at a heating rate of 40 K/min. The average grain size determined by TEM, the mean crystallite size and the dislocation density obtained from x-ray line profile analysis and the onset temperature of recovery/recrystallisation measured by DSC are listed in Table 7.1 [12]. The twin boundary frequency for all samples was relatively low (0.1 ± 0.1%), close to the detection limit of this quantity for the applied experimental setup of x-ray line profile analysis. The very high dislocation density after HPT can be attributed to the high pressure (p = 4 GPa) applied during HPT, which hinders vacancy migration, thereby impeding the climb-controlled annihilation of dislocations. Table 7.1 reveals that despite the similar grain size values of the samples processed by various SPD methods, the onset temperatures of recovery/recrystallisation are very different. It can be established that the higher the dislocation density and lower the crystallite size, the lower the onset temperature of recovery/recrystallisation, i.e. the samples produced by MDF and TE show the highest thermal stability while the HPT-processed specimen has the lowest stability. It is noted that the very high dislocation density after HPT is most probably accompanied by very large long-range internal stresses. These stresses may assist annihilation of dislocations, thereby resulting in lower temperature of recovery/recrystallisation, as has already been suggested in previous papers [e.g. 13]. It is also noted that in addition to the grain size and the dislocation density, other features of the microstructure (e.g. grain boundary character) may also influence the stability of the SPD-processed samples. Electron backscatter diffraction (EBSD) analysis (not shown here) revealed that the fraction of high angle grain boundaries has similar high values (80–94%) for all the samples presented in Table 7.1, therefore most probably it does not increase significantly the difference in stability.

Table 7.2 shows a comparison between the parameters of the microstructure and the thermal stability of nanostructured

Table 7.1 **The average grain size determined by TEM, the mean crystallite size, the dislocation density and the twin boundary frequency obtained from x-ray line profile analysis and the onset temperature of recovery/recrystallisation (T_{onset}) measured by DSC at a heating rate of 40 K/min for Cu processed by different SPD methods**

Processing method	Grain size [nm]	Crystallite size [nm]	Dislocation density [10^{14} m^{-2}]	T_{onset} [K]
20 MDF	225	142±15	7±1	581±3
15 TE	225	107±12	10±1	572±3
25 ECAP	215	101±15	15±1	469±2
25 HPT	160	75±9	37±4	419±2

Source: Reprinted from Gubicza et al. (2011) [12] with permission from Elsevier.

Ni processed by different methods including ECAP, cold rolling (CR), HPT, their combinations and electrodeposition [14]. The details of processing methods are given in Zhilyaev et al. (2003) [15]. For the SPD-processed Ni samples, the thermal stability decreases when different SPD procedures are combined due to the reduction in grain size and the increase of dislocation density. At the same time, the electrodeposited Ni has higher stability than the samples processed by SPD despite the much smaller grain size and much higher dislocation density in the former material. Most probably, this effect is caused by the higher impurity level in Ni sample produced by electrodeposition.

In commercially pure hexagonal UFG materials the annihilation of dislocations starts before the appearance of exothermic DSC peak, as illustrated for Ti processed by 8 ECAP passes in Figure 7.3, where the mean crystallite size and the dislocation density as a function of annealing

Table 7.2 The average grain size determined by TEM, the mean crystallite size and the dislocation density obtained from x-ray line profile analysis and the peak temperature of recovery/recrystallisation (T_{peak}) measured by DSC at a heating rate of 40 K/min for Ni processed by different SPD methods

Processing method	Grain size [nm]	Crystallite size [nm]	Dislocation density [10^{14} m^{-2}]	T_{peak} [K]
8 ECAP	350	71±5	9±1	673
8 ECAP + 85%CR	300	38±4	11±1	647
5 HPT	170	42±4	17±2	542
8 ECAP + 5 HPT	140	48±4	25±2	570
8 ECAP + 85%CR + 5 HPT	100	46±3	20±2	563
Electrodeposition	35	24±3	82±4	610

Source: The data are taken from Zhilyaev et al. (2002) [14].

temperature are shown [3]. At 800 K, where the DSC peak starts, the dislocation density has already decreased from 28 × 10^{14} m^{-2} to 6.6 × 10^{14} m^{-2}. The increase of the crystallite size is accelerated only after 800 K and it reaches 164 nm at 850 K. The dislocation density decreases further after 800 K and at 850 K it has a value of 5 × 10^{13} m^{-2}. The change of the prevailing dislocation slip systems in the sample was also determined as a function of temperature by x-ray line profile analysis. The procedure is described in the Appendix. The 11 dislocation slip systems can be classified into three groups based on their Burgers vectors: $b_1 = 1/3\langle 11\bar{2}0\rangle$ (<*a*> type), $b_2 = \langle 0001\rangle$ (<*c*> type) and $b_3 = 1/3\langle 11\bar{2}3\rangle$ (<*c*+*a*> type). In the ECAP-processed Ti sample the relative fractions of <*a*>, <*c*> and <*c*+*a*> Burgers vectors are 64 ± 2%, 2 ± 2% and 34 ± 2%,

Figure 7.3 **The mean crystallite size and the dislocation density as a function of annealing temperature for Ti processed by 8 ECAP passes. The corresponding DSC thermogram is also presented**

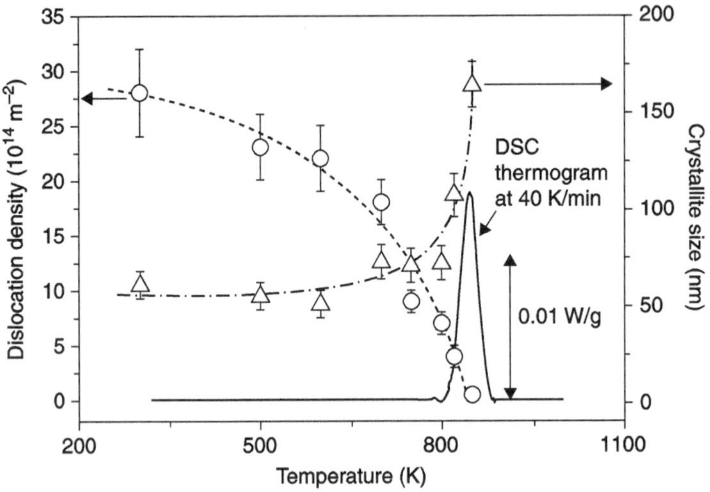

respectively. The abundance of <*a*>-type dislocations besides <*c*>- and <*c*+*a*>-type dislocations can be explained by their smallest Burgers vector. Figure 7.4 shows that as the temperature increases, the relative fraction of <*c*+*a*>-type dislocations decreases down to about 2%, indicating that these dislocations disappear faster than <*a*> or <*c*>-type ones [3]. This can be explained by the fact that <*c*+*a*>-type dislocations have larger Burgers vector and consequently higher formation energy than the other two types.

The activation energy of recovery/recrystallisation can be determined from the shift of the exothermic DSC peak corresponding to the variation of heating rate using the Kissinger equation [16]:

$$\ln\frac{\beta}{T_p^2} = -\frac{Q}{R}\frac{1}{T_p} + A, \qquad (7.1)$$

Figure 7.4 The relative fractions of <*a*>-, <*c*>- and <*c*+*a*>-type dislocations as a function of annealing temperature

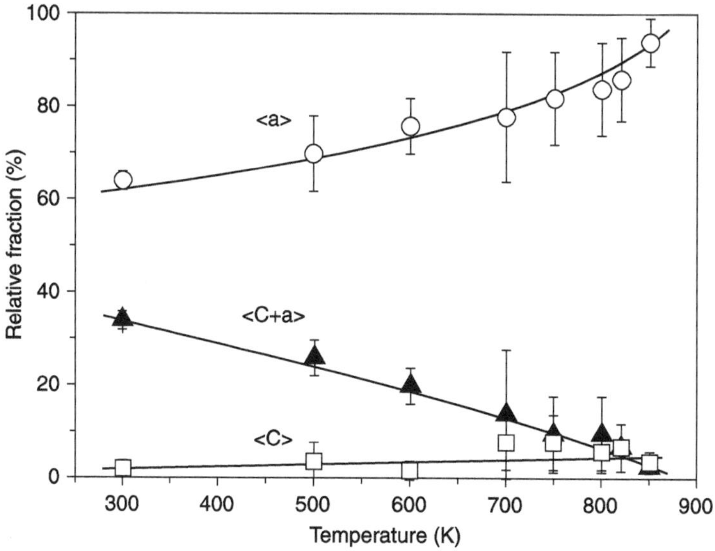

where β is the heating rate, T_p is the peak temperature, Q is the activation energy, R is the universal gas constant, and A is a constant. It is noted that at atmospheric pressure the activation energy is very close to the activation enthalpy. The activation energies determined for various pure UFG and nanomaterials along with their processing method and grain size are listed in Table 7.3. The grain size was determined by TEM except for the Cu sample consolidated by cold compaction of sputtered nanocrystalline particles. In that case, the crystallite size was obtained by x-ray diffraction line profile analysis. The heat released in the exothermic peak corresponding to recovery/recrystallisation is also presented in the table as this value usually does not depend on the heating rate. It can be concluded that for the majority of materials the activation energy is about $0.5 \pm 0.1 \times Q_{self}$, where Q_{self} is the activation energy of self-diffusion,

Table 7.3 The processing method, the grain size, the heat released in the exothermic peak and the activation energy of recrystallisation determined for various UFG and nanomaterials. The grain size was determined by TEM except for the Cu sample consolidated by cold compaction of sputtered nanocrystalline particles. In that case, the crystallite size was obtained by x-ray diffraction line profile analysis (XLPA)

Material	Processing	Grain size [nm]	Released heat [J/g]	Activation energy of recrystallisation [kJ/mol]	Ratio of the activation energy of recrystallisation and the bulk diffusion activation energy	Reference
Cu 99.98 wt.%	ECAP, 5 passes	300	1.39	121	0.61	[17]
Cu 99.98 wt.%	ECAP, 8 passes	250	0.88	119	0.60	[17]
Cu OFHC	ECAP, 8 passes	210	0.95	80	0.4	[18]
Cu 98.51 wt.% – Si 1.49 wt.%	HPT, 5 GPa, 1 revolution	470	1.12	94	0.47	[1]
Cu 99.99 wt.%	HPT, 4 GPa, 2 revolutions	–	1.27	75	0.37	[13]
Cu 99.99 wt.%	HPT, 3 GPa, 1 revolution	150	–	96	0.48	[19]
Cu	Cold compaction of sputtered particles	9 (XLPA)	23	100	0.50	[20]
Cu	Uniaxial compaction of inert gas condensed particles	40	5	83	0.42	[21]

Ni 99.99 wt.%	ECAP, 8 passes	350	1.0	108	0.39	[14, 22, 23]
Ni 99.99 wt.%	ECAP, 8 passes + cold rolling (85%)	300	1.27	106	0.38	[14]
Ni 99.99 wt.%	HPT, 6 GPa, 5 revolutions	170	1.75	89	0.32	[14]
Ni 99.998 wt.%	HPT, 4 GPa, 2 revolutions	–	2.5	91	0.33	[13]
Ni 99.99 wt.%	ECAP, 8 passes + HPT, 6 GPa, 5 revolutions	140	3.19	99	0.35	[14]
Ni 99.99 wt.%	ECAP, 8 passes + cold rolling + (85%)HPT, 6 GPa, 5 revolutions	100	3.88	98	0.35	[14]
Ni	electrodeposition	35	10.3	108	0.39	[14]
Ni	electrodeposition	20	7	132	0.47	[24]
Ag 99.995%	ECAP, 16 passes	200	0.8	85	0.46	Not published
Ag	Uniaxial compaction of inert gas condensed particles	60	3	101	0.55	[25]
hcp Co	electrodeposition	20	–	151	0.54	[26]

irrespective of the type of material and the processing method. Previous studies have shown that the activation energy of diffusion along grain boundaries and dislocations is also about half of that for self-diffusion [27–30]. Therefore, the observed values of activation energy of recovery/recrystallisation in nanomaterials can be explained by the large amount of dislocations and grain boundaries, which can act as diffusion paths during annealing. The data in Table 7.3 reveal that the grain size has no significant effect on the activation energy. Figure 7.2b also shows that the activation energy for Ag remains unchanged with increasing number of ECAP passes, i.e. its value is less sensitive to the actual defect structure. At the same time, the released heat increases steeply with decreasing grain size below 200 nm, as illustrated in Figure 7.5. It should be noted that in addition

Figure 7.5 **The released heat obtained in DSC experiments as a function of grain size**

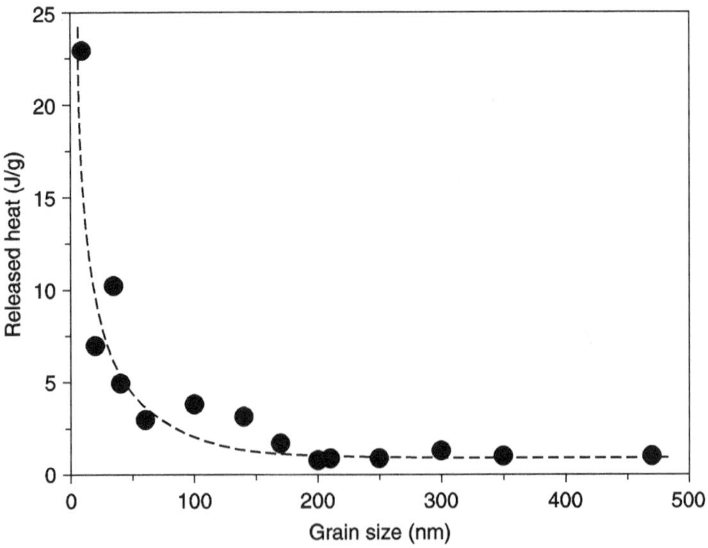

Source: The data are taken from Table 7.3.

to grain boundaries, most probably vacancies and dislocations also contribute to the released heat.

In the case of 99.998% and 99.99% purity Ni samples processed by HPT at RT, two exothermic peaks were detected on the DSC thermograms [6, 13]. The first peak appearing at lower temperature (about 120°C at 10 K/min) corresponds to the disappearance of excess monovacancies, while the larger second one (at about 275°C) is associated with the annihilation of vacancy clusters and dislocations as well as recrystallisation. The first peak was not observed for 99.99% Cu processed under the same conditions as most probably all the vacancies are clustered [6]. In the case of Ni, the high values of stacking fault energy (SFE) and melting point, and the large pressure during HPT retard the clustering of vacancies. The heat released in the first peak is about one order of magnitude smaller than for the second peak. The concentration of vacancies (CV) annihilated during the first peak has been determined from the released heat (H) using the formula [6]:

$$H = e_{\mathrm{vac}} c_{\mathrm{v}} \frac{N_A}{M_{\mathrm{mol}}} \tag{7.2}$$

where e_{vac} is the formation energy of a vacancy (0.29×10^{-18} J for Ni), N_{A} is the Avogadro's number (6×10^{23} mol^{-1}) and M_{mol} is the molar mass for Ni (58.93 g·mol^{-1}). The vacancy concentration was found to be about 10^{-4}, which is close to the equilibrium value at the melting point. With increasing the number of HPT revolutions, the temperature of the first peak remained more or less constant while the second one shifted to lower temperatures, similarly as shown in Figure 7.1. This difference can be attributed to the short- and long-range strain fields of vacancies and dislocations, as they decrease as r^{-3} and r^{-1}, respectively, where r is the distance from the defect. With increasing the number of HPT revolutions, the dislocation density increases, resulting in higher interaction stresses between them, which facilitates their annihilation and shifts the DSC peak to lower temperature. Due to the short-range strain field around vacancies, this effect is negligible when their concentration increases, therefore the temperature of the first DSC peak does not change significantly with increasing the number of HPT revolutions [6, 13]. The activation energies evaluated by the Kissinger analysis are 62 and 91 kJ/mol for the processes occurring in the first (monovacancy annihilation) and second (vacancy cluster and dislocation annihilation as well as recrystallisation) peaks, respectively. Neither the number of HPT revolutions nor the impurity level affected significantly the values of activation energies.

7.2 Stability of nanostructured Cu during storage at room temperature

Nanostructured materials have non-equilibrium microstructures with a high stored enthalpy due to the large amount of lattice defects (vacancies, dislocations, planar faults and grain boundaries). The high enthalpy acts as a driving force in annihilation of lattice defects, i.e. reduction in vacancy concentration, dislocation density and grain-growth. Despite the high driving force for recovery and recrystallisation, the nanostructures formed in commercially pure materials are usually stable at the temperature of processing due to the retarding effect of impurities and alloying elements. At the same time, in pure nanomaterials annihilation of lattice defects may occur during their storage at the processing temperature (usually at RT). For instance, in the case of 99.98% purity Cu processed by ECAP, the area under the exothermic DSC peak was reduced by a factor of 2–3 during the storage for four years at RT, indicating a decrease of the energy stored in the microstructure [31]. Figure 7.6 shows DSC thermograms taken at 40 K/min heating rate immediately after ECAP and storage for four years at RT for Cu samples processed by 1 and 10 passes. A similar phenomenon was observed for the samples processed by 3 and 25 passes.

The reduction in the stored energy indicates the annihilation of lattice defects. After four years of storage the parameters of the microstructure determinable by x-ray line profile analysis (crystallite size, dislocation density, edge/screw character and arrangement of dislocations and twin boundary

Figure 7.6 DSC thermograms obtained immediately after ECAP and storage at RT for four years in the case of Cu samples processed by 1 and 10 passes

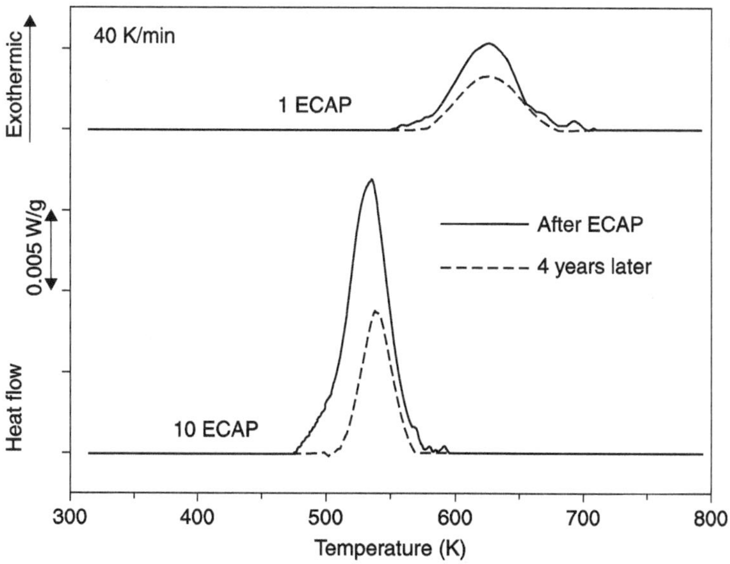

Source: Reprinted from Gubicza et al. (2010) [31] with permission from Elsevier.

frequency) remained unchanged within the experimental error, which indicates either that recrystallisation did not occur or that it is marginal during storage for four years. If a significant amount of new, dislocation-free grains had been formed, the average dislocation density would have decreased. It should be noted that annihilation of dislocations and recrystallisation may occur during longer storage of these Cu samples similarly as reported in previous studies, e.g. in Mishin and Godfrey (2008) [32]. The results of x-ray line profile analysis show that the reduction in the stored energy during storage is most probably attributed to the annihilation of lattice defects, which cannot be evaluated by line profile analysis. These defects may be point defects such as vacancies or vacancy agglomerates which form in a high concentration during severe plastic deformation, as shown in previous papers [5, 13, 33]. The relation between the stored energy of vacancies per unit mass and the vacancy concentration (c_v) is given by Equation 7.2, where e_{vac} and M_{mol} are 0.195×10^{-18} J [6] and 63.55 g mol^{-1}, respectively, for Cu. Assuming that the reduction in the stored energy is attributed solely to the loss of vacancies, the decrease of vacancy concentration during the storage for four years was determined from the change of stored energy using Equation 7.2. The change of c_v is plotted as a function of number of ECAP passes in Figure 7.7. Earlier studies [6] have shown that in the case of SPD-processed UFG Cu, the vacancies agglomerated into clusters during SPD, therefore most probably mainly clustered vacancies are annihilated during storage at RT.

The loss of vacancy concentration (~1–3 × 10^{-4}) has similar value to the vacancy content determined previously for Cu samples immediately after ECAP by electrical resistivity measurement or a combination of x-ray line profile analysis and calorimetry (see also Chapter 2) [5]. Therefore, most probably the majority of vacancies formed during ECAP

Figure 7.7 **The reduction in vacancy concentration determined from the decrease of stored energy in Cu as a function of number of ECAP passes**

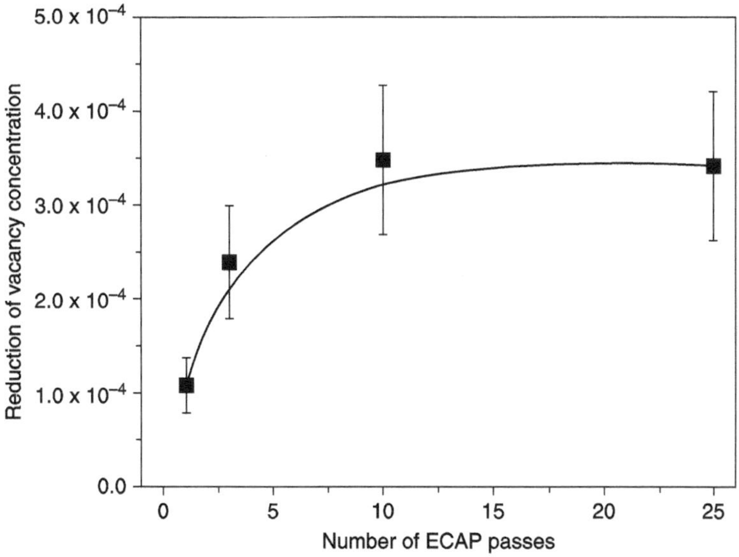

Source: Reprinted from Gubicza et al. (2010) [31] with permission from Elsevier.

disappeared during storage for four years at RT. To support this assumption, the stored energy of dislocations (E_{disl}) was determined from the dislocation density listed in Table 7.4 using the following formula [5]:

$$E_{\text{disl}} = A\frac{Gb^2\rho}{\rho_{\text{m}}}\ln\frac{1}{b\sqrt{\rho}}, \qquad (7.3)$$

where G is the shear modulus (47 GPa), b is the magnitude of Burgers vector (0.25 nm), ρ is the dislocation density, ρ_{m} is the mass density (8.96×10^6 gm^{-3} for Cu) and A stands for the factor depending on the edge/screw character of dislocations. The value of A equals to $(4\pi)^{-1}$ and $(4\pi(1-\nu))^{-1}$ for screw and edge dislocations, respectively, where ν is the Poisson's ratio (0.3 was taken). The parameter q determined from x-ray line profile analysis describes the edge/screw character of dislocations. The theoretically calculated values of q for pure edge and screw dislocations in Cu are 1.68 and 2.37, respectively [34]. The values of q obtained for the samples processed by different number of ECAP passes are listed in Table 7.4. In the case of mixed dislocations the value of A can be obtained from the experimentally determined q using a simple rule of mixture:

$$A = \frac{q-1.68}{0.69}\frac{1}{4\pi} + \frac{2.37-q}{0.69}\frac{1}{4\pi(1-\nu)}. \tag{7.4}$$

The values of stored energy of dislocations (E_{disl}) calculated from Equations 7.3 and 7.4 are listed in Table 7.4. The values of E_{disl} are smaller than the released heat measured after four years (H_{4y}). The difference (H_{4y} – E_{disl}) can be attributed to the grain boundaries and the remaining vacancies. As this difference is 4–9 times smaller than the stored energy of vacancies annihilated during four years ($H_{ECAP} - H_{4y}$), therefore the remaining vacancy concentration should be much smaller than that of the vacancies disappeared during four years, i.e. the values of c_v plotted in Figure 7.7 approximate well the concentrations of vacancies formed during ECAP.

It is worth noting that the values of c_v obtained in this evaluation have the same order of magnitude as the equilibrium vacancy concentration near the melting point. Taking 1.28 eV as the value of vacancy formation energy and 12.2 as the pre-exponential factor in the Arrhenius-type formula of vacancy concentration [35], 4×10^{-21} and 2×10^{-4} are obtained for the equilibrium vacancy concentrations at RT and the melting point, respectively. This means that the vacancy concentration after ECAP at RT is 17 orders of magnitude larger than its equilibrium value. The very large excess of vacancies was most probably formed due to severe

Table 7.4 **The crystallite size, the dislocation density (ρ) and the parameter q describing the edge/screw character of dislocations obtained by x-ray line profile analysis. The released heat obtained immediately after ECAP (H_{ECAP}) and storage for four years (H_{4y}) by DSC, and the contribution of dislocations (E_{disl}) to the stored energy calculated from Equations 7.3 and 7.4**

ECAP passes	Crystallite size [nm]	ρ [10^{14} m^{-2}]	q	H_{ECAP} [J/g]	H_{4y} [J/g]	E_{disl} [J/g]
1	87±9	15±2	2.0±0.1	0.48±0.04	0.28±0.03	0.23±0.03
3	75±8	21±2	2.0±0.1	0.85±0.09	0.41±0.04	0.31±0.04
10	64±7	20±2	1.9±0.1	1.0±0.1	0.38±0.04	0.31±0.04
25	101±10	15±2	1.7±0.1	1.0±0.1	0.37±0.04	0.27±0.04

Source: Reprinted from Gubicza et al. (2010) [31] with permission from Elsevier.

plastic deformation during ECAP. During storage of the ECAP-processed samples, the vacancy concentration might decrease by migration along the boundaries of UFG grains to the free surface and/or by sinking at small dislocation loops. In the latter case, the sink of vacancies is accompanied by the annihilation of dislocation loops. These small loops having a dimension of about 1–2 nm may be formed when mobile dislocations intersect immobile dislocation dipoles. These loops are also invisible in x-ray line profile analysis due to their strongly shielded strain fields and small dimensions. If a large number of loops act as sinks of vacancies, the annihilation of these loops also contributes to the decrease of the stored energy during storage of the samples, therefore the loss of vacancies may be lower than the values plotted in Figure 7.7. It is noted that the large reduction in vacancy concentration did not yield significant change of the hardness (1.28 GPa) for the ECAP-processed sample during storage for four years.

The UFG 99.98% purity Cu processed by 25 HPT revolutions exhibited even stronger recovery during storage at RT, which is accompanied by softening [12]. Generally, the phenomenon when a nanostructure softens due to recovery and recrystallisation during storage at the temperature of processing is called 'self-annealing'. Figures 7.8a and 7.8b show TEM images taken immediately after HPT-processing of Cu and four years of storage, respectively. The comparison of the images revealed a growth of the mean grain size from about 160 nm to 250 nm. It is noted that some relatively large grains with the size of about 600 nm were also observed after storage for four years. X-ray line profile analysis showed that the average crystallite size increased from 75 ± 9 to 122 ± 15 nm while the dislocation density decreased from $37 \pm 4 \times 10^{14}$ to $13 \pm 2 \times 10^{14}$ m^{-2} during storage for four years. Simultaneously, the twin

Figure 7.8 **TEM images taken immediately after HPT-processing (a) and four years of storage (b)**

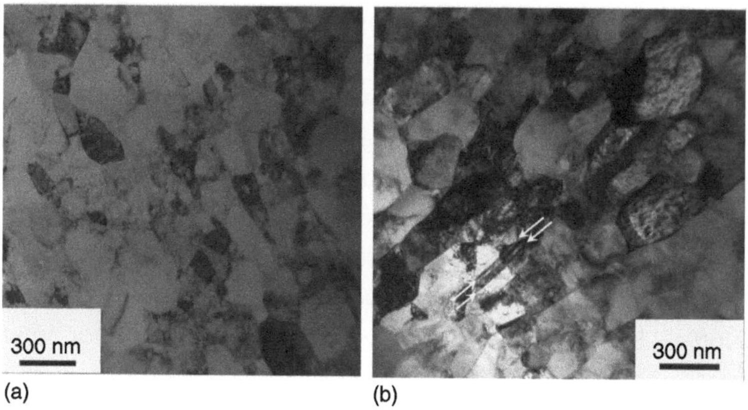

Source: Reprinted from Gubicza et al. (2011) [12] with permission from Elsevier.

boundary frequency increased from 0.1 ± 0.1% to 0.3 ± 0.1%. In the TEM image of Figure 7.8b some twin boundaries are indicated by white arrows. These changes in the microstructure indicate recovery and/or recrystallisation during storage of the sample. As was discussed in Chapter 2, the high pressure applied during HPT hinders the annihilation processes of dislocations resulting in very high dislocation density and small grain size. During storage of the sample, this extremely high dislocation density and the very small grain size are the driving forces of the microstructure transformation. The high density of dislocations results in large internal stresses. Therefore, when the pressure was released after HPT, the vacancy diffusion became faster, resulting in annihilation of edge dislocations by climb. Moreover, the internal stresses most probably also cause cross-slip of dissociated screw dislocations after HPT. The dislocation density can also decrease by twinning at dislocation pile-ups developed at glide obstacles. If the local stresses at these obstacles exceed the critical stress required for twin nucleation, after an incubation time twins are formed at the expense of dislocations in the pile-ups (see Chapter 3) [36]. The operation of this mechanism is supported by the increase of twin boundary frequency from $0.1 \pm 0.1\%$ to $0.3 \pm 0.1\%$ during storage. It is noted, however, that partial recrystallisation may also contribute to the higher twin boundary frequency as twinning usually occurs during recrystallisation of face-centred cubic (fcc) metals. The hardness of the HPT-processed sample decreased from 1.69 ± 0.06 GPa to 1.07 ± 0.12 GPa during storage for four years as a result of recovery/recrystallisation of the microstructure.

It is noted that self-annealing was also observed previously in ECAP-processed Cu at RT [32, 37, 38]. For example, large recrystallised grains were observed in 99.96% purity Cu eight years after processing by ECAP [32]. Experiments on 5N purity Cu showed that recrystallisation occurred only two months after ECAP and the lower recrystallisation time for the purer material was explained by the non-availability of the pinning effect of alloying elements on grain boundaries and dislocations [37]. Nanocrystalline Cu and Ag samples processed by the consolidation of nanopowders also show self-annealing at RT during periods of several days after their production [39, 40]. Additionally, it was shown for nanocrystalline Cu and Ag layers, having a thickness of ~1–20 μm and processed by electroplating, that recrystallisation occurs at RT within ~1–2 days of their production [41–44]. It was suggested that during layer processing the organic additives pin the grain boundaries, thereby stabilising the small grain size [41, 42]. However, the additives disappear from the layers within several hours after plating and this leads to recrystallisation of the nanostructures.

7.3 Self-annealing in nanostructured silver: the significance of a very low stacking fault energy

The low SFE in pure materials decreases the stability of nanostructures. First, the smaller the SFE, the larger the distance between partials in dissociated dislocations. The high degree of dislocation dissociation hinders the annihilation by either cross-slip or climb, which leads to a very large dislocation density during grain refinement by SPD (see Chapter 3). Secondly, the low SFE is associated with a small value of twin boundary energy. For instance, in fcc structures the twin boundary energy is about half of SFE. The low SFE facilitates the formation of recrystallised grains as the nuclei can be separated by low-energy twin boundaries from the parent grains. The latter effect gives a smaller stability of nanostructures prepared by either SPD-processing methods or procedures without SPD. It is noted that the low stability of nanomaterials having small SFE can be effectively increased by alloying.

Silver has the lowest SFE among fcc metals, therefore the self-annealing was studied on high-purity UFG 99.995% Ag processed by ECAP [45]. Silver billets were pressed through totals of 1, 4, 8 and 16 passes of ECAP at RT with a pressing velocity of 8 $mm \cdot s^{-1}$. Following ECAP, the billets were stored at RT and the microstructures and mechanical behaviour were examined as a function of the time of storage for periods up to a total of four months. Figure 7.9 shows the microhardness of samples processed by different numbers of ECAP passes as a function of the time of storage at RT, where the lower horizontal line denotes the hardness of the initial sample. It is apparent that the hardness after 1 pass remains unchanged within experimental error even after storage for four months. By contrast, the hardness gradually decreases

Figure 7.9 The microhardness of samples processed by different numbers of ECAP passes as a function of the time of storage at RT

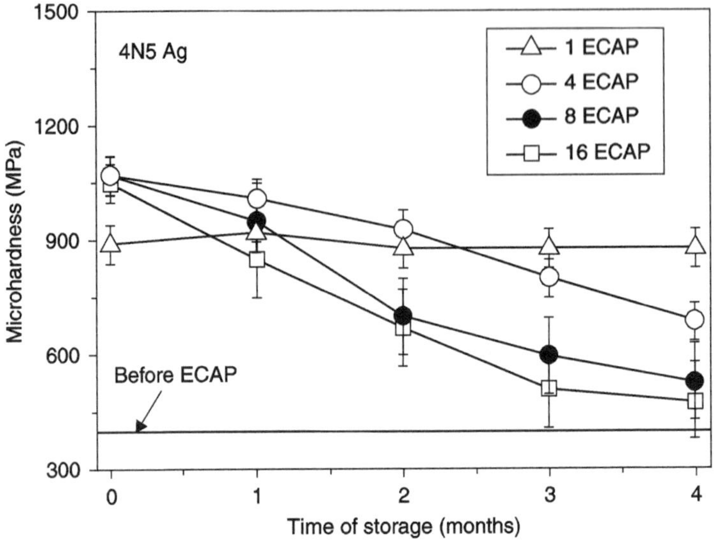

Source: Reprinted from Gubicza et al. (2010) [45] with permission from Elsevier.

with increasing storage time for the samples processed by 4, 8 and 16 passes, thereby demonstrating that the severely deformed microstructure was self-annealed during long-term storage at RT. Compression experiments also confirmed the softening of the samples during storage [45].

Complementary x-ray diffraction line profile analysis and TEM investigations revealed that the softening in UFG Ag during storage is caused by both recovery and recrystallisation [45]. Figure 7.10a shows a Debye-Scherrer diffraction ring of the 220 reflection obtained immediately after 8 passes of ECAP. In this condition the intensity distribution around the ring is relatively homogeneous, thereby indicating a high degree of homogeneity within the microstructure. After four months of storage at RT high intensity spots were observed on the Debye-Scherrer rings for the samples processed through 4, 8 and 16 passes. Figure 7.10b shows an example for the 220 reflection obtained after processing through 8 passes and storing for four months. The inhomogeneous intensity distributions in the Debye-Scherrer rings suggest there is a fluctuation in the degree of self-annealing within the microstructure. Since the breadths of the individual high intensity spots were close to the instrumental broadening, it was not feasible to evaluate these parts of the diffraction rings quantitatively. However, these high intensity spots are most probably scattered from recrystallised grains having larger sizes and lower dislocation densities than the detection limits of the line profile analysis (~800 nm and ~10^{13} m^{-2} for the crystallite size and the dislocation density, respectively) and the rather homogeneous portions of the Debye-Scherrer rings are most probably related to the recovered volumes. The

Figure 7.10 Debye-Scherrer rings for the 220 reflection of x-rays (a) immediately after 8 ECAP passes and (b) after 8 ECAP passes and storage at RT for four months

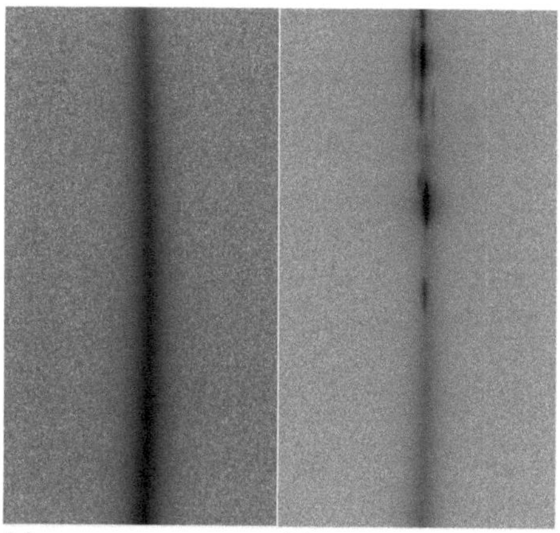

(a) (b)

Source: Reprinted from Gubicza et al. (2010) [45] with permission from Elsevier.

comparison of the TEM images taken immediately after 8 passes and storage for four months in Figure 7.11 proves the occurrence of recrystallisation [45]. The ECAP-processed sample has an average grain size of about 200 nm while in the recrystallised volumes the size increased to ~1 μm. The Debye-Scherrer rings for the samples processed for 4 and 16 passes also became inhomogeneous during storage at RT, whereas the diffraction rings remained homogeneous in the sample processed by 1 pass even after storage for four months.

The dislocation density and the twin boundary frequency evaluated from the homogeneous portions of the rings are shown as a function of the storage time in Figures 7.12a and 7.12b, respectively [45]. For the sample processed by 1 pass the dislocation density in Figure 7.12a remains essentially unchanged whereas after 4, 8 and 16 passes the dislocation density decreases with the time at different rates. The reduction rate in the dislocation density increases in the order of 16, 4

Figure 7.11 **Bright field TEM images from a sample processed through 8 ECAP passes (a) and after storage at RT for four months (b)**

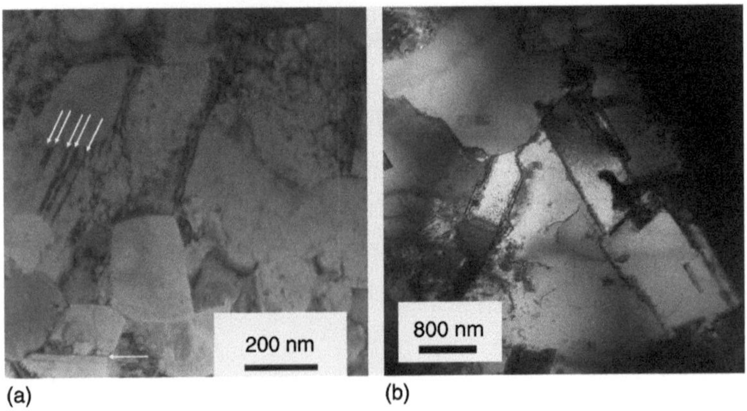

Source: Reprinted from Gubicza et al. (2010) [45] with permission from Elsevier.

Figure 7.12 (a) The dislocation density and (b) the twin boundary frequency in the recovered volumes of samples stored at RT up to four months after processing by ECAP through 1, 4, 8 and 16 passes

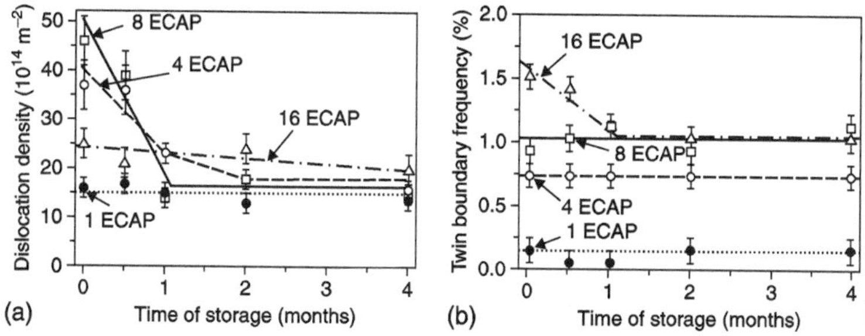

Source: Reprinted from Gubicza et al. (2010) [45] with permission from Elsevier.

and 8 passes of ECAP, which follows the trend in the initial dislocation density determined immediately after ECAP-processing. The higher the initial dislocation density, the faster the recovery. With increasing time of storage, the dislocation density for each specimen converges to the value characteristic of the sample pressed through only 1 pass (~16×10^{14} m^{-2}). Figure 7.12b shows that the twin boundary frequency remains practically unchanged for specimens processed through 1, 4 and 8 passes, but for the 16 passes sample the twin boundary frequency decreases from ~1.5 ± 0.1% to ~1.1 ± 0.1% during the first month of storage and thereafter remains unchanged. It is emphasised that the defect densities plotted in Figure 7.12 were determined from the homogeneous parts of the Debye-Scherrer rings, which are free from the large intensity spots. Consequently, these values are characteristic only of the recovered parts of the samples and most probably the mean dislocation density and twin boundary frequency are lower for the whole specimens because of contributions from the recrystallised fractions of the samples.

The recovery of the severely deformed microstructure in Ag may occur by cross-slip and/or climb of dislocations. It is well known that both mechanisms are thermally activated and stress-assisted processes and the probability of their occurrence decreases with increasing degree of dislocation dissociation [46, 47]. The large splitting distance between partials in Ag leads to a low probability of cross-slip and climb, which initially impedes the annihilation of dislocations during ECAP processing, thereby resulting in very high dislocation density, but this process may occur by thermal activation after sufficiently long times. In the absence of any assisting stresses, the waiting time for cross-slip in Ag is of the order of ~10^{13} years, which means in practice that cross-slip cannot occur in Ag without assisting stresses (see Chapter 3). In a severely deformed microstructure, however, there are significant internal stresses, originating primarily from the very high density of dislocations, and these stresses will assist the cross-slip and climb processes. For example, assuming a value for the assisting stresses of ~10 MPa, the waiting time for cross-slip in silver is reduced to ~10^7 s. This means in practice that cross-slip may occur some months after the SPD processing, thereby giving a delayed recovery for the dislocation structures in Ag processed by ECAP in agreement with the experimental observations. The inhomogeneous recovery of the dislocation structure is attributed to the stress sensitivity of the probability of cross-slip and climb [47]. In severely deformed polycrystals, the magnitude of the remaining stresses acting on the glide planes after deformation will change from grain to grain due to the distributions of the dislocations and other defects, such as twins, so that the driving force for cross-slip and climb will depend upon the specific location within the specimen. As a result, it is anticipated that there will be fluctuations in the waiting times for annihilation of dislocations and therefore the recovery starts inhomogeneously within the sample. It is apparent from Figure 7.12a that higher dislocation densities after ECAP are associated with faster rates of recovery so that shorter times are then needed to achieve the stable dislocation density during storage at RT. For all ECAP samples, the stable dislocation density is close to the value obtained immediately after 1 pass ($\sim 16 \times 10^{14}$ m^{-2}). The higher initial dislocation density corresponds to a shorter mean spacing between dislocations and therefore higher assisting stresses and consequently faster annihilation. For the sample processed by 1 pass, the lower dislocation density explains the lack of any significant recovery and also the termination in recovery in the samples processed by 4 and 8 passes when the dislocation density reaches the value characteristic for 1 pass. It should be noted that an apparent decrease of the dislocation density in the recovered but non-recrystallised fraction of the samples may also occur if the recrystallisation starts in the volumes having the highest dislocation density.

X-ray line profile analysis shows that in the recovered volumes of samples processed through 4 to 16 passes, the dislocation density and the twin boundary frequency do not decrease below the value characteristic of the sample pressed through 1 pass (Figure 7.12), while the hardness (Figure 7.9) for the samples for 4 to 16 passes fall below the values for 1 pass after storage for four months. This apparent dichotomy is readily explained by the partial recrystallisation in the samples processed through 4 to 16 passes. By increasing the numbers of passes up to 8, the mean dislocation density measured immediately after ECAP increases. The higher dislocation densities in the non-recovered volumes of the samples will lead to higher driving forces for recrystallisation and accordingly the rate of recrystallisation should be faster after larger numbers of passes. At the same time, the mean dislocation density measured immediately after 16 passes is smaller than after 8 passes, while after four months the hardness reduction is the highest for the sample processed by 16 passes. This is attributed to the higher degree of recrystallisation in this sample. Processing by ECAP between 8 and 16 passes gives an increase in the twin boundary frequency together with a decrease in the total dislocation density. In locations where the twin boundary frequency increases at the expense of dislocations, the stored energy will decrease as the twin fault energy in Ag is very low (~8 mJ m^{-2} [48]). These volumes will then act as the nuclei in recrystallisation and therefore the time required for grain nucleation will be the shortest for the sample processed by 16 passes. In this case the strongly twinned regions are recrystallised first, therefore twin boundary frequency decreases in the recovered but non-recrystallised volumes for the sample processed by 16 passes, which is in line with the experimental results. This is also consistent with observations that the nuclei of recrystallised grains form primarily at deformation twins in low SFE metals, such as Ag [49].

The formation of regions where twinning occurs at the expense of dislocations will increase the degree of inhomogeneity in the microstructure and introduce fluctuations in the stored energy in the sample processed by 16 passes. This is consistent with other studies, suggesting that heterogeneities in the microstructure may promote recrystallisation due to the higher strain energy gradients [37, 50]. It is noted that the lower grain boundary energy associated with the larger grain size in the sample processed by only 1 pass (~30 μm) than in the specimens processed by 4 to 16 passes (~200 nm) may also contribute to the lack of recrystallisation in the former material. It should also be noticed that a recent study of annealing in Cu processed by ECAP showed that recrystallisation starts heterogeneously primarily in regions where grain boundaries have high angles of misorientation [51]. Since the angle of misorientation at grain boundaries usually increases with increasing numbers of passes in ECAP [52, 53], it is reasonable to anticipate that recrystallisation in Ag will be faster after larger numbers of passes. It is noted that self-annealing was also observed in HPT-processed high purity Ag (99.99%) and Au (99.999%) samples [54].

7.4 References

1. H. Jiang, Y.T. Zhu, D.P. Butt, I.V. Alexandrov and T.C. Lowe, 'Microstructural evolution, microhardness and thermal stability of HPT-processed Cu', *Materials Science and Engineering*, A 290 (2000) 128–138.
2. N. Takata, K. Yamada, K, Ikeda, F. Yoshida, H. Nakashima and N. Tsuji, 'Annealing Behavior and Recrystallized Texture in ARB Processed Copper', *Materials Science Forum*, 503–504 (2006) 919–924.
3. J. Gubicza, N.H. Nam, L. Balogh, R.J. Hellmig, V.V. Stolyarov, Y. Estrin and T. Ungár, 'Microstructure of severely deformed metals determined by X-ray peak profile analysis', *Journal of Alloys and Compounds*, 378 (2004) 248–252.
4. J. Gubicza, L. Balogh, R. J. Hellmig, Y. Estrin, T. Ungár, 'Dislocation structure and crystallite size in severely deformed copper by X-ray peak profile analysis' *Materials Science and Engineering*, A 400–401 (2005) 334–338.
5. E. Schafler, G. Steiner, E. Korznikova, M. Kerber, M. J. Zehetbauer, 'Lattice defect investigation of ECAP-Cu by means of X-ray line profile analysis, calorimetry and electrical resistometry' *Materials Science and Engineering*, A 410–411 (2005) 169–173.

6. D. Setman, E. Schafler, E. Korznikova, M.J. Zehetbauer, 'The presence and nature of vacancy type defects in nanometals detained by severe plastic deformation' *Materials Science and Engineering*, A 493 (2008) 116–122.
7. N. Gao, M.J. Starink and T.G. Langdon, 'Using differential scanning calorimetry as an analytical tool for ultrafine grained metals processed by severe plastic deformation', *Materials Science and Technology*, 25 (2009) 687–698.
8. F. Dalla Torre, R. Lapovok, J. Sandlin, P.F. Thomson, C.H.J. Davies and E.V. Pereloma 'Microstructures and properties of copper processed by equal channel angular extrusion for 1–16 passes', *Acta Materialia*, 52 (2004) 4819–4832.
9. J. Gubicza, N. Q. Chinh, J.L. Lábár, S. Dobatkin, Z. Hegedűs and T.G. Langdon, 'Correlation between microstructure and mechanical properties of severely deformed metals', *Journal of Alloys and Compounds*, 483 (2009) 271–274.
10. Y.M. Wang and E. Ma, 'Three strategies to achieve uniform tensile deformation in a nanostructured metal', *Acta Materialia*, 52 (2004) 1699–1709.
11. R.Z. Valiev, 'Nanomaterial advantage', *Nature*, 419 (2002) 887–889.
12. J. Gubicza, S.V. Dobatkin, E. Khosravi, A.A. Kuznetsov and J.L. Lábár, 'Microstructural stability of Cu processed by different routes of severe plastic deformation', *Materials Science and Engineering*, A 528 (2011) 1828–1832.
13. D. Setman, M.B. Kerber, E. Schafler and M.J. Zehetbauer, 'Activation Enthalpies of Deformation-Induced Lattice Defects in Severe Plastic Deformation Nanometals Measured by Differential Scanning Calorimetry', *Metallurgical and Materials Transactions*, A 41 (2010) 810–815.

14. A.P. Zhilyaev, G.V. Nurislamova, S. Surinach, M.D. Baró and T.G. Langdon, 'Calorimetric measurements of grain growth in ultrafine-grained nickel', *Material Physics and Mechanics*, 5 (2002) 23–30.
15. A.P. Zhilyaev, J. Gubicza, G. Nurislamova, Á. Révész, S. Suriñach, M.D. Baró and T. Ungár, 'Microstructural characterization of ultrafine-grained nickel', *Physica Status Solidi*, (a) 198 (2003) 263–271.
16. H.E. Kissinger, 'Reaction kinetics in differential thermal analysis', *Analytical Chemistry*, 29 (1957) 1702–1706.
17. N. Lugo, N. Llorca, J.J. Sunol and J.M. Cabrera, 'Thermal stability of ultrafine grains size of pure copper obtained by equal-channel angular pressing', *Journal of Materials Science*, 45 (2010) 2264–2273.
18. W.Q. Cao, C.F. Gu, E.V. Pereloma and C.H.J. Davies, 'Stored energy, vacancies and thermal stability of ultra-fine grained copper', *Materials Science and Engineering*, A 492 (2008) 74–79.
19. J. Cizek, I. Prochazka, M. Cieslar, R. Kuzel, J. Kuriplach, F. Chmelik, I. Stulikova, F. Becvar and O. Melikhova, 'Thermal stability of ultrafine grained copper', *Physical Review*, B 65 (2002) 094106.
20. Y.K. Huang, A.A. Menovsky and F.R. de Boer, 'Calorimetric analysis of the grain growth in nanocrystalline copper samples', *Nanostructured Materials*, 2 (1993) 587–595.
21. A. Kumpmann, B. Günther and H.-D. Kunze, 'Thermal stability of ultrafine-grained metals and alloys', *Materials Science and Engineering*, A168 (1993) 165–169.
22. A.P. Zhilyaev, B.-K. Kim, J.A. Szpunar, M.D. Baro and T.G. Langdon, 'The microstructural characteristics of ultrafine-grained nickel', *Materials Science and Engineering*, A 391 (2005) 377–389.

23. A.P. Zhilyaev, G.V. Nurislamova, R.Z. Valiev, M.D. Baro and T.G. Langdon, 'Thermal Stability and Microstructural Evolution in Ultrafine-Grained Nickel after Equal-Channel Angular Pressing (ECAP)', *Metallurgical and Materials Transactions*, A 33 (2002) 1865–1868.
24. N. Wang, Z. Wang, K.T. Aust and U. Erb, 'Isokinetic analysis of nanocrystalline nickel electrodeposits upon annealing', *Acta Materialia*, 45 (1997) 1655–1669.
25. A. Kumpmann, B. Günther and H.-D. Kunze, 'Thermal stability of ultrafine-grained metals and alloys', *Materials Science and Engineering*, A168 (1993) 165–169.
26. G. Hibbard, K.T. Aust, G. Palumbo and U. Erb, 'Thermal stability of electrodeposited nanocrystalline cobalt', *Scripta Materialia*, 44 (2001) 513–518.
27. J. Lian, R.Z. Valiev and B. Baudelet, 'On the enhanced grain growth in ultrafine grained metals', *Acta Metallurgica, et Materialia*, 43 (1995) 4165–4170.
28. S.C. Tjong and H. Chen, 'Nanocrystalline materials and coatings', *Materials Science and Engineering*, R 45 (2004) 1–88.
29. C-M. Kuo and C-S. Lin, 'Static recovery activation energy of pure copper at room temperature', *Scripta Materialia*, 57 (2007) 667–670.
30. S. Divinski, J. Ribbe, G. Schmitz and C. Herzig, 'Grain boundary diffusion and segregation of Ni in Cu', *Acta Materialia*, 55 (2007) 3337–3346.
31. J. Gubicza, S.V. Dobatkin and E. Khosravi, 'Microstructural stability of Cu processed by different routes of severe plastic deformation', *Materials Science and Engineering*, A 527 (2010) 6102–6104.
32. O.V. Mishin and A. Godfrey, 'Microstructure of ECAE-processed copper after long-term room-temperature storage', *Metallurgical and Materials Transactions*, A 39A (2008) 2923–2930.

33. T. Ungár, E. Schafler, P. Hanák, S. Bernstorff and M. Zehetbauer, 'Vacancy production during plastic deformation in copper determined by in situ X-ray diffraction', *Materials Science and Engineering*, A 462 (2007) 398–401.
34. T. Ungár, J. Gubicza, G. Ribárik and A. Borbély, 'Crystallite size distribution and dislocation structure determined by diffraction profile analysis: principles and practical application to cubic and hexagonal crystals', *Journal of Applied Crystallography*, 34 (2001) 298–310.
35. P. Varotsos and K. Eftaxias, 'High-temperature vacancy concentration in Cu', *Physical Review*, B 40 (1989) 9963–9964.
36. J.B. Cohen and J. Weertman, 'A dislocation model for twinning in fcc metals', *Acta Metallurgica*, 11 (1963) 996–998.
37. G. Wang, S.D. Wu, L. Zuo, C. Esling, Z.G. Wang and G.Y. Li, 'Microstructure, texture, grain boundaries in recrystallization regions in pure Cu ECAE samples', *Materials Science and Engineering*, A 346 (2003) 83–90.
38. Y. Estrin, N.V. Isaev, S.V. Lubenets, S.V. Malykhin, A.T. Pugachov, V.V. Pustovalov, E.N. Reshetnyak, V.S. Fomenko, L.S. Fomenko, S.E. Shumilin, M. Janecek and R.J. Hellmig, 'Effect of microstructure on plastic deformation of Cu at low homologous temperatures', *Acta Materialia*, 54 (2006) 5581–5590.
39. B. Gunther, A. Kumpmann and H.-D. Kunze, 'Secondary recrystallization effects in nanostructured elemental metals', *Scripta Metallurgica et Materialia*, 27 (1992) 833–838.
40. V.Y. Gertsman and R. Birringer, 'On the structure and strength of ultrafine-grained copper produced by severe plastic deformation', *Scripta Metallurgica et Materialia*, 30 (1994) 577–581.

41. K.B. Yin, Y.D. Xia, C.Y. Chan, W.Q. Zhang, Q.J. Wang, X.N. Zhao, A.D. Li, Z.G. Liu, M.W. Bayes and K.W. Yee, 'The kinetics and mechanism of room-temperature microstructural evolution in electroplated copper foils', *Scripta Materialia*, 58 (2008) 65–68.
42. M. Stangl, M. Lipták, A. Fletcher, J. Acker, J. Thomas, H. Wendrock, S. Oswald and K. Wetzig, 'Influence of initial microstructure and impurities on Cu room-temperature recrystallization (self-annealing)', *Microelectronic Engineering*, 85 (2008) 534–541.
43. K. Hansen and K. Pantleon, 'Microstructure stability of silver electrodeposits at room temperature', *Scripta Materialia*, 58 (2008) 96–98.
44. J.-M. Paik, Y.-J. Park, M.-S. Yoon, J.-H. Lee and Y.-C. Joo, 'Anisotropy of grain boundary energies as cause of abnormal grain growth in electroplated copper films', *Scripta Materialia*, 48 (2003) 683–688.
45. J. Gubicza, N.Q. Chinh, J.L. Lábár, Z. Hegedűs and T.G. Langdon, 'Principles of self-annealing in silver processed by equal-channel angular pressing: The significance of a very low stacking fault energy', *Materials Science and Engineering*, A. 527 (2010) 752–760.
46. A.S. Argon and W.C. Moffatt, 'Climb of extended edge dislocations', *Acta Metallurgica*, 29 (1981) 293–299.
47. P.B. Escaig, 'Sur le glissement dévié des dislocations dans la structure cubique à faces centrées', *Journal de Physique*, 29 (1968) 225–239.
48. J.P. Hirth and J. Lothe, *Theory of Dislocations*, Wiley, New York (1982).
49. H. Paul, J.H. Driver, C. Maurice and A. Piatkowski, 'The role of shear banding on deformation texture in low stacking fault energy metals as characterized on model Ag crystals', *Acta Materialia*, 55 (2007) 575–588.

50. M. Kumar, A.J. Schwartz and W.E. King, 'Microstructural evolution during grain boundary engineering of low to medium stacking fault energy fcc materials', *Acta Materialia*, 50 (2002) 2599–2612.
51. O.V. Mishin and J.R. Bowen, 'Through-Thickness Variations of Deformed and Annealed Microstructures in ECAE-Processed Copper', *Metallurgical and Materials Transactions*, A 40 (2009) 1684–1692.
52. M. Kawasaki, Z. Horita and T.G. Langdon, 'Microstructural evolution in high purity aluminum processed by ECAP', *Materials Science and Engineering*, A 524 (2009) 143–150.
53. S.D. Terhune, D.L. Swisher, K. Oh-ishi, Z. Horita, T.G. Langdon and T.R. McNelley, 'An investigation of microstructure and grain-boundary evolution during ECA pressing of pure aluminum', *Metallurgical and Materials Transactions*, 33A (2002) 2173–2184.
54. H. Matsunaga and Z. Horita, 'Softening and Microstructural Coarsening without Twin Formation in FCC Metals with Low Stacking Fault Energy after Processing by High-Pressure Torsion', *Materials Transactions*, 50 (2009) 1633–1637.

8

Relationship between microstructure and hydrogen storage properties of nanomaterials

Abstract: The influence of particle size, crystallite size and lattice defects on hydrogen storage capacity and absorption-desorption kinetics of nanostructured materials is overviewed. It is shown that the particle and crystallite sizes, and catalyst play two different roles in the sorption process. By decreasing the particle and crystallite sizes, the required diffusion path of hydrogen is drastically reduced, which enhances the sorption kinetics significantly. At the same time, catalysts have a promoting effect on dissociation or recombination of hydrogen molecules on the particles' surfaces and they also act as chemisorption sites. The lattice defects (dislocations, stacking faults and twin boundaries) facilitate the diffusion of hydrogen and the nucleation of hydride phases. It is revealed that large amounts of dislocations and/or planar faults are formed due to stresses induced by phase transformations in hydrogenation and dehydrogenation processes. The effect of defects on hydrogen storage capacity of carbon nanotubes is also reviewed.

Key words: hydrogen storage, absorption, desorption, contracting volume model, dislocation, stacking fault, twin boundary, catalyst.

8.1 Fundamentals of hydrogen storage in solid state materials

The application of hydrogen as a fuel can help to overcome the challenges arising from the fossil fuel-based economy, such as the release of growing amounts of greenhouse gas CO_2, urban air quality and the reduction in the world crude oil supply. Hydrogen can be produced directly from sunlight and water by biological organisms and using semiconductor-based systems similar to photovoltaics, or indirectly via thermal processing of biomass or fossil fuels [1]. The electric power necessary for the decomposition of water into hydrogen and oxygen can be produced by wind turbines or nuclear power plants during off-peak periods. In a hydrogen fuel cell, electricity and water are produced in the following manner. Hydrogen molecules diffuse through a porous anode towards a catalyst made of Pt and are stripped of their electrons and become positively charged ions (protons). The anode and the cathode are separated by a proton-permeable polymeric membrane that allows protons to migrate to the cathode [1]. The electrons formed at the anode-side catalyst cannot migrate through the polymeric membrane, but rather travel in the external circuit between the anode and the cathode, thereby creating electric current. At the cathode side, on the surface of a catalyst the hydrogen protons recombine with electrons and oxygen molecules in air to produce water and heat. This waste heat gives the fuel cell an operating temperature of 60–80°C.

Hydrogen should be stored for supplying the fuel cells. A possible solution is the storage of hydrogen in a chemically bonded state, forming solid-state hydrides in metals or intermetallic compounds. Because of the low pressures involved in metal hydride technologies and the fact that the release of hydrogen takes place via an endothermic process, this method of hydrogen storage is very safe. Additionally, the volumetric hydrogen capacity, defined as the mass of hydrogen in a unit volume of the storing device, is about 80–160 kg/m^3 for solid-state hydrides, which is much higher than that either for compressed hydrogen gas under 80 MPa pressure (~40 kg/m^3) or for liquid hydrogen in a tank at -252°C (~71 kg/m^3). During the solid-state absorption process, first hydrogen molecules are adsorbed onto the surface and dissociate into atoms by breaking the molecular bonds in favour of new bonds to the surface at chemisorption sites [1]. The energy to overcome the activation potential of dissociation is usually supplied by the vibrational energy. Then, the hydrogen atoms enter the material by diffusion. Following diffusion of hydrogen atoms into the bulk lattice sites, a hydride is formed by nucleation and growth. The formation of hydride starts at the surface of the metal, therefore hydrogen atoms should migrate through the hydride to continue hydrogenation in the metal. In the case of desorption, metal has to be nucleated at the surface, and hydrogen atoms have to diffuse to the surface and recombine to hydrogen molecules, which have to physically desorb [2]. In this case the hydrogen atoms should diffuse through the metal formed at the surface of grains [3].

The absorption-desorption behaviour can be characterised by the pressure–composition isotherm depicted schematically in Figure 8.1. This curve can be obtained by measuring the hydrogen pressure outside the hydrogen storing material as a function of hydrogen concentration in weight per cent in the hydrogen storing material during absorption and desorption at a given temperature. The uptake of hydrogen occurs at higher pressure than desorption, which yields a hysteresis in the curve. The origin of hysteresis is mainly attributed to the excess of energy required to override the lattice strains occurring upon hydrogen absorption or to the production of misfit dislocations between the initial and the hydride phases. The metal initially dissolves only a small amount of hydrogen, which creates a solid solution of hydrogen in the metal matrix [1]. As the hydrogen pressure and hydrogen concentration in the metal are increasing, interactions between hydrogen and metal atoms become locally important and nucleation and growth of a new metal hydride phase is observed. In the plateau region, a mixture of solid solution and metal hydride phases exists. The length of plateau determines the reversible capacity of the material, which gives the relative weight fraction of hydrogen that can be stored reversibly with a small pressure variation (see Figure 8.1). The desorption temperatures for high-capacity hydrides at a pressure of 1 bar are between 110°C and 600°C. At this atmospheric pressure, the reversible capacity for these materials varies between 5 and 14 wt.%. With increasing

Figure 8.1 Schematic diagram of pressure–composition isotherm

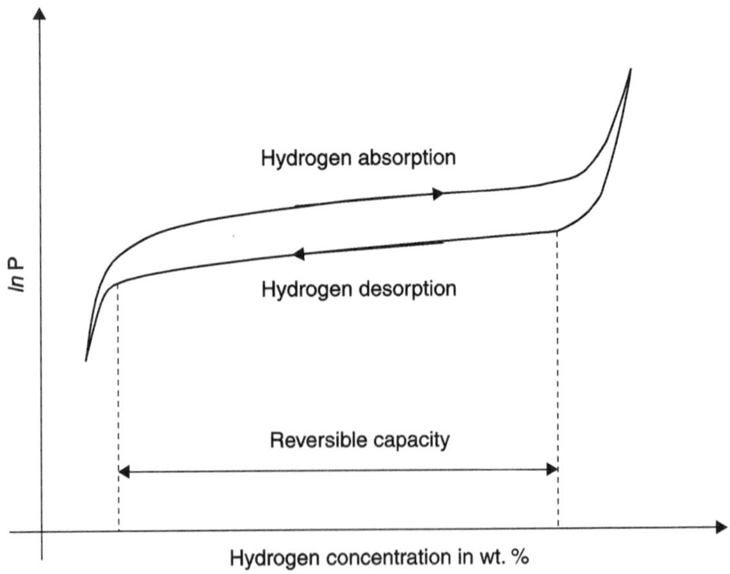

temperature, the plateau pressure increases and beyond a critical temperature, the plateau region disappears. The relation between the pressure P measured at the middle of the absorption plateau and the temperature T is given by the Van't Hoff equation [1]:

$$\ln\left(\frac{P}{P_0}\right)=\frac{\Delta H}{RT}-\frac{\Delta S}{R}, \tag{8.1}$$

where P_0 is the atmospheric pressure, ΔH and ΔS are the enthalpy and entropy changes during the hydriding reaction, respectively, T is the absolute temperature, and R is the universal gas constant. For almost all hydrides the enthalpy and entropy changes during hydriding reaction are negative, i.e. the hydriding reaction is exothermic and dehydriding reaction is endothermic. The entropy change has nearly constant value of $\Delta S \approx -130\ \mathrm{J \cdot mol^{-1} \cdot K^{-1}}$ for all the solid state hydrogen systems while ΔH varies between –20 and –200 kJ/mol for different materials [1].

Basically, there are three models describing absorption/desorption kinetics: (i) surface reaction; (ii) Johnson-Mehl-Avrami-Kolmogorov (JMAK); and (iii) contracting volume (CV) models of phase transformations [4]. The surface reaction model assumes that the slowest step of reaction is the chemisorption, i.e. the dissociation or recombination of hydrogen molecules on the particles' surfaces. In this case, the transformed fraction, V_{rel} linearly depends on time t as:

$$V_{\mathrm{rel}}=k_{CS}t, \tag{8.2}$$

where k_{CS} is the reaction constant. In JMAK theory of phase transformations, it is assumed that the nucleation and growth of the new (hydride) phase starts randomly in the bulk and at the surface. The transformed fraction as a function of time can be given as:

$$V_{rel} = 1 - \exp\left[-(k_{JMAK}t)^n\right], \tag{8.3}$$

where k_{JMAK} is the reaction constant and n depends on the dimensionality of the growth of the new phase nuclei (e.g. n=2 corresponds to two-dimensional growth and n=3 to three-dimensional growth). Assuming that the hydrogen diffusion is very fast, the transformation rate is controlled by the (constant) velocity of the metal/hydride interface [4].

If the nucleation of the hydride phase starts at the surface of the metal particle and the growth continues from the surface into the bulk, the contracting volume (CV) model is used. The analytical model assumes that a thin layer of transformed phase on the surface of the particle already exists. The kinetics can be described as:

$$V_{rel} = 1 - (1 - k_{CV}t)^n, \tag{8.4}$$

where n depends again on the dimensionality of the growth ($n = 2$ and 3 for two- and three-dimensional growth processes, respectively). The reaction constant k_{CV} is proportional to the velocity of the metal/hydride interface and inversely proportional to the radius of the metal particle. If the hydrogen diffusion is fast enough, the rate of transformation is determined by the constant velocity of the interface. The difference in growth mechanism for JMAK and CV models is illustrated in Figure 8.2 [4].

If the diffusion of hydrogen through the transformed phase is the rate-limiting step, the interface velocity decreases with time. In case of contracting volume and three-dimensional growth, the relationship between the transformed volume fraction and the time can be given as:

$$1 - \left(\frac{2V_{rel}}{3}\right) - (1 - V_{rel})^{2/3} = k_{DCV} \cdot t, \tag{8.5}$$

Figure 8.2 Schematic depiction of phase transformation according to (a) Johnson-Mehl-Avrami-Kolmogorov and (b) contracting volume models. The dark areas represent the growing new phase

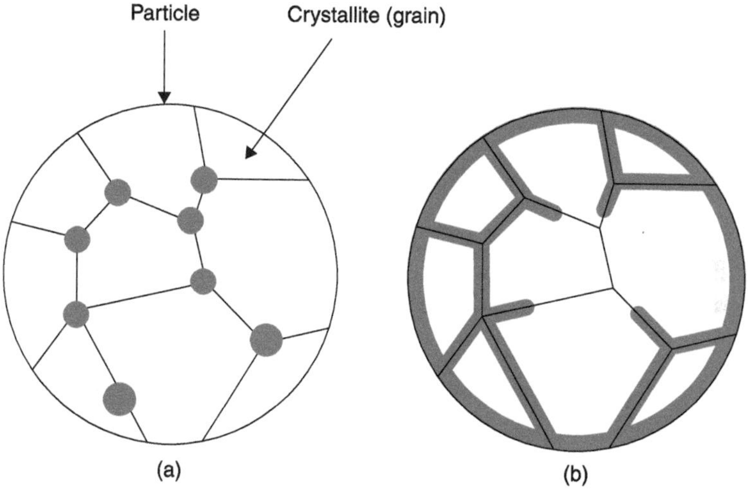

where k_{DCV} is the reaction constant [4]. The comparison of the experimental V_{rel} versus t data with Equations 8.2–8.5 may yield an estimation of the rate-controlling mechanism of hydrogen absorption and desorption, as will be shown in the next section.

8.2 Microstructure and hydrogen storage in nanomaterials processed by severe plastic deformation

During the absorption process, hydrogen atoms first migrate along the grain boundary network and then they begin to diffuse into grain interiors. Therefore, the nanostructured state of materials used for hydrogen storage is beneficial from the viewpoint of fast absorption and desorption of hydrogen. The nanocrystalline microstructure is usually achieved by severe plastic deformation, such as ball milling of powders or high-pressure torsion (HPT) of bulk materials. During milling, both the powder particle size and the grain size inside the particles decrease, leading to a faster diffusion of hydrogen along the particle surfaces and the grain boundaries. Additionally, grain boundaries are favourable nucleation sites for the formation and decomposition of the hydride phase. Moreover, high-energy milling (i) breaks the oxide layers and passivation coatings on particles' surfaces, hence exposure of fresh catalytic sites for dissociation of hydrogen molecules, (ii) increases the densities of lattice defects (e.g. dislocations, stacking faults and twin boundaries) that also serve as paths for fast diffusion of hydrogen [1]. Moreover, addition of catalysts to hydriding material further increases the rate of sorption by assisting the dissociation of hydrogen molecules. Thus, the three dominant factors influencing the sorption properties are particle size, grain size and catalyst.

The effects of grain and particle sizes on hydrogen storage properties were studied extensively on milled MgH_2 powders. It was shown in Chapter 4 that the grain size determined by electron microscopy and the crystallite size obtained by x-ray line profile analysis broadly coincide for ball-milled materials, therefore the two terms will be used equivalently in this chapter. Magnesium is considered as one of the most attractive hydrogen storage materials, mainly because of high storage capacity (7.6 wt.%), lightweight and low cost [1]. During hydrogenation, Mg having hexagonal crystal structure transforms to tetragonal MgH_2. Hydrogen atoms bind too strongly with the Mg atoms, i.e. the absolute value of formation enthalpy of MgH_2 is high (–75 kJ per mol H_2) [1], therefore the hydride needs to be heated up to very high desorption temperatures (~350°C) in order to release hydrogen gas at atmospheric pressure (~1 bar). Additionally, the absorption and desorption are also very slow at temperatures ~350°C, which impedes the direct use of magnesium as hydrogen storage material. To convert Mg completely to MgH_2 requires more than 50 hours at 350°C [5]. The slow sorption kinetics at lower temperatures are mainly due to the low dissociation ability of hydrogen gas molecules on the metallic Mg surface and the slow diffusion of hydrogen through the hydride formed on the surface of Mg particles. Ball-milling of Mg in hydrogen atmosphere yields a nanocrystalline material with the crystallite size of ~30 nm, which can absorb about 6 wt.% hydrogen in much shorter time, 120 minutes, and at lower temperature, 300°C, than in the case of coarse-grained Mg. Milling MgH_2 instead of pure Mg results in even smaller crystallite size of ~10 nm, which yields very short duration, 10 minutes, of sorption at 300°C. The addition of suitable catalysts (e.g. Nb_2O_5) leads to very short hydrogen absorption and desorption times of less than 2 minutes at 300°C for MgH_2 with high capacity of 7 wt.% hydrogen [6]. The milling of MgH_2 is usually carried out in hydrogen atmosphere in order to prevent the decomposition of MgH_2 into Mg and H_2. The effects of particle and crystallite sizes, and the catalyst on the absorption and desorption behaviours at 300°C, are illustrated in Figure 8.3 [2]. The absorption and desorption are performed under 8 bar hydrogen and in vacuum, respectively. The conventional MgH_2 powder with coarse particles (>10 μm) and crystallites (>1 μm) cannot be loaded or unloaded with hydrogen at 300°C in reasonable times. The powders processed by ceramic ball-milling for 20 hours and 700 hours have comparably small crystallite sizes in the range of 5–20 nm, while the particle sizes are different, ~5 and ~0.5 μm, respectively. This difference results in a much faster sorption kinetics for the sample milled for 700 hours. It should be noted that the reduction of grain and particle sizes also decrease the hydrogen desorption temperature [7].

The addition of Nb_2O_5 catalyst to MgH_2 powder before milling increases further the rate of absorption and desorption of hydrogen (see Figure 8.3). However, the particle/crystallite size and catalyst play two different roles in the sorption process. By decreasing the particle and crystallite sizes, the required diffusion path of hydrogen is drastically reduced, which enhances the sorption kinetics significantly. However, the size reduction influences neither the adsorption of hydrogen molecules on the surface, nor the ease of their dissociation. At the same time, the metal-oxide catalysts seem to have a promoting effect on dissociation or recombination of hydrogen molecules on the particles' surfaces and they also act as chemisorption sites [6], resulting in a decrease of the activation energy of hydrogenation and

Figure 8.3 Absorption (left) and desorption curves (right) of (a) conventional coarse-grained MgH_2; (b) nanocrystalline MgH_2 processed by ball-milling for 20 hours; (c) MgH_2 ball-milled for 700 hours and (d) Nb_2O_5-catalyzed ball-milled MgH_2

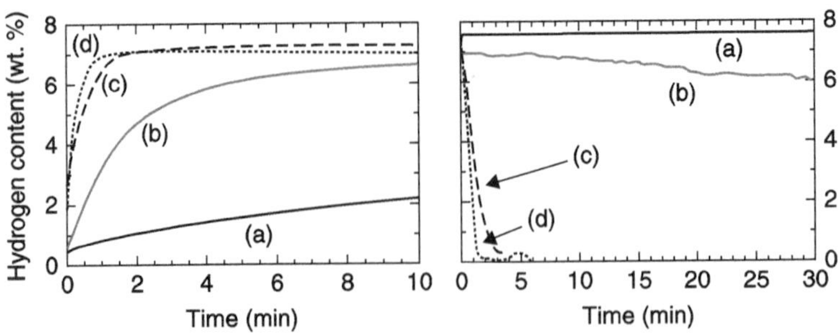

Source: Reprinted from Dornheim et al. (2007) [2] with permission from Elsevier.

dehydrogenation [4, 7]. Moreover, the hard Nb_2O_5 catalyst particles in the milled powder blend results in an additional slight decrease in MgH_2 particle size, which also contributes to better sorption kinetics [7]. It seems that above a critical fraction, additional catalyst does not yield further acceleration of absorption or desorption. In the case of Nb_2O_5 catalyst added to MgH_2, this critical amount is ~0.5 mol% when hydrogenation and dehydrogenation are performed under hydrogen atmosphere at 8.4 bar and in vacuum, respectively [4]. After 100 hours' milling, the activation energy of desorption is drastically decreased from values around 120 kJ/mol for the chemisorption of uncatalysed nanocrystalline magnesium hydride down to about 62 kJ/mol by increasing the catalyst content up to ~0.5 mol%. Above this catalyst content, further decrease of activation energy is not observed. As the activation energy of desorption for MgH_2 with coarse particles is ~140 kJ/mol, the reduction of particle and crystallite sizes in MgH_2 during milling yields only a slight decrease in the activation energy.

By comparing the experimental absorption and desorption data with Equations 8.2–8.5 for MgH_2 powder milled with Nb_2O_5 catalyst, the absorption behaviour is best described by the three-dimensional diffusion-controlled CV model (Equation 8.5) [4]. In the dehydrogenation process, generally chemisorption (recombination of hydrogen molecules on the surface) is the rate-controlling mechanism (Equation 8.2), but when both milling time and catalyst content are high, CV model with two-dimensional growth (Equation 8.4, $n=2$) best describes the kinetics. Both the increase of milling time – leading to the reduction of particle and crystallite sizes – and Nb_2O_5 content increase the desorption reaction constants, k_{CS} and k_{CV} for chemisorption and CV model, respectively [4]. These results indicate that during desorption the hydrogen diffusion through the dehydrided Mg is fast and not rate-limiting. On the other hand, the absorption of hydrogen in Mg is mainly controlled by the diffusion of hydrogen through the hydride phase formed at the surface of the particles. This difference between the absorption and desorption behaviours can be explained by the smaller diffusion constant of hydrogen through MgH_2 compared with Mg [8]. The increment in the reaction constant for absorption with increasing the catalyst content and the milling time can be explained by an enhanced dissociation rate of hydrogen molecules resulting in a higher gradient of the hydrogen chemical potential and thereby in a faster diffusion of hydrogen through the formed magnesium hydride phase [4]. Longer milling reduces the particle and crystallite sizes, yielding a faster diffusion of hydrogen along boundaries. When a very high fraction of catalysts is added and the long milling time yields a nanosized dispersion of catalyst particles, the recombination of hydrogen atoms to molecules at the surface is fast and no longer the rate-limiting step in the desorption process. In this case, the powder particles are covered by a dehydrided magnesium phase and the interface between the dehydrided magnesium and the magnesium hydride moves from the particles' surfaces along the crystallite boundary areas into the volume of the powder particles because the interface velocity along crystallite boundaries is significantly faster than perpendicular to the crystallite boundary area. As a consequence, the desorption kinetics is governed by the interface mobility of the transformed phase along the two-dimensional network of crystallite boundaries, which is reflected by the value $n=2$ in CV model [4]. It is noted that taking the size distribution of crystallites into account, the kinetics and the reaction constants of the different transformation models are considerably modified compared with the models for monodisperse crystallites [9].

The severe plastic deformation of bulk Mg and its alloys also improves the hydrogen storage properties [10–16]. For instance, in 3N purity Mg processed by HPT up to 10 revolutions under a pressure of 6 GPa total hydrogen absorption of 6.9 wt.% is achieved at 150°C under a hydrogen pressure of 3 MPa. The dislocation density reaches its maximum value after 1/4 revolution, however, after 10 turns both the dislocation density and the grain size are reduced due to primary recrystallisation. Although the dislocation density is much larger in the sample processed up to 1/4 revolution than after 10 turns, the hydrogen storage capacity of the former material is comparably small (<0.5 wt.%) than the annealed Mg with very large grain size of ~1.6 mm [16]. After 10 revolutions, absorbed hydrogen is increased to 6.9 wt.% and the hydrogen absorption rate is 12 times faster than for the annealed sample or the sample after 1/4 turn, which can be attributed to significant grain refinement and an increase in the area fraction of high-angle grain boundaries [16]. It is concluded that in HPT-processed Mg high-angle grain boundaries play a more effective role than dislocations in the hydrogen absorption. The improvement of hydrogen storage properties is also observed for bulk Mg-Ti alloys processed by accumulative roll-bonding [17].

The effect of stacking faults and twin boundaries on hydrogen storage properties is investigated in only a few studies [1, 18–20]. For instance, high density of stacking faults is formed during ball milling of battery-grade $Ni(OH)_2$ used in the positive electrodes of rechargeable alkaline nickel metal hydride batteries. $Ni(OH)_2$ has a hexagonal structure with well-stacked basal planes that are bonded by weak Van der Walls forces. Ball-milling of $Ni(OH)_2$ yields stacking faults on basal planes, which facilitates accommodation of hydrogen ions between basal planes.

8.3 Change of defect structure during dehydrogenation–hydrogenation cycles

The dehydrogenation–hydrogenation cycle leads to changes in the microstructure of hydrogen storage nanomaterials and therefore a variation of their absorption/desorption kinetics [21]. Figure 8.4 shows the first and the second desorption curves measured in vacuum for MgH_2 ball-milled for 10 hours at 300°C. The first dehydrogenation process needs longer time (~80 minutes) than the second one (~10 minutes). For subsequent cycles, further changes in the desorption curves are not observed. The longer dehydrogenation time in the first cycle can be attributed to the non-activated surface of the particles in the as-milled powder. In the activation process occurring during the first cycle, Mg nuclei precipitate on surfaces of the as-milled MgH_2 particles below the

Figure 8.4 **The first and the second desorption curves measured in vacuum at 300 °C for MgH_2 ball-milled for 10 hours. The first desorption corresponds to the activation of the powder**

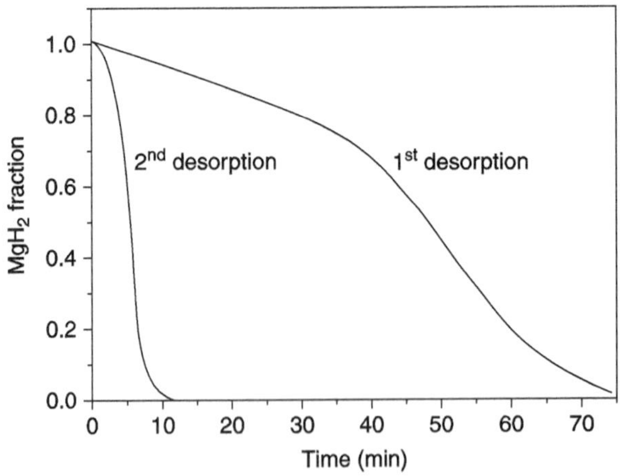

Source: The data are taken from Fátay et al. (2007) [22].

magnesium-oxide and hydroxide passivation layer. As the nuclei grow, the induced strains cause the fracture of the passivation layer, exposing a fresh activated surface [22]. After the first cycle, further desorption processes take shorter times due to the activated surfaces of particles. In contrast to the desorption behaviour, the first absorption curve is not significantly different from the subsequent ones, indicating that no further activation of the powder occurs during the absorption interval of the first full cycle (the cycle starts with desorption). The absorption was measured under a hydrogen pressure of 8 bar.

The MgH_2 sample milled for 10 hours at 300°C has a mean crystallite and particle sizes of 9 nm and 780 nm, respectively. The variation of the average crystallite size of MgH_2 as a function of the number of desorption–absorption cycles is shown in Figure 8.5. During the first full dehydrogenation–

Figure 8.5 Variation of the average crystallite size of MgH_2 as a function of the number of dehydrogenation-hydrogenation cycles

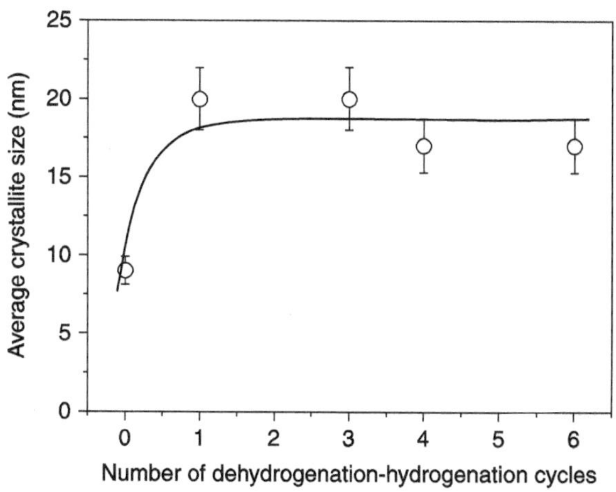

Source: Reprinted from Fátay et al. (2007) [22] with permission from Elsevier.

hydrogenation cycle at 300°C, the average crystallite size increases from 9 nm to 20 nm, while the crystallite-size distribution becomes narrower, i.e. the crystallite structure becomes more homogeneous [22]. Further dehydriding/ hydriding treatments increase the width of size distribution, while the average crystallite size remains unchanged. After six cycles the crystallite size distribution is slightly wider than for the as-milled MgH_2 powder. The crystallite structures in the as-milled MgH_2, and after the first and the second desorption–absorption cycles are illustrated in Figure 8.6. The crystallite-growth in MgH_2 during the first cycle is caused by the high temperature and the relatively long time of desorption required for the first full cycle at 300°C. After activation, the desorption occurs on a considerably shorter time scale, therefore some residual MgH_2 inclusions may remain inside the larger Mg crystallites, since the transformed Mg spreads from the boundaries toward the interiors of crystallites according to CV model (see section 8.1). The relative volume fraction of the possible residual MgH_2 inclusions has no detectable contribution to the measured

Figure 8.6 **Schematic depiction of the crystallite structure in MgH_2 (a) immediately after milling, after (b) the first and (c) the second dehydrogenation–hydrogenation cycles**

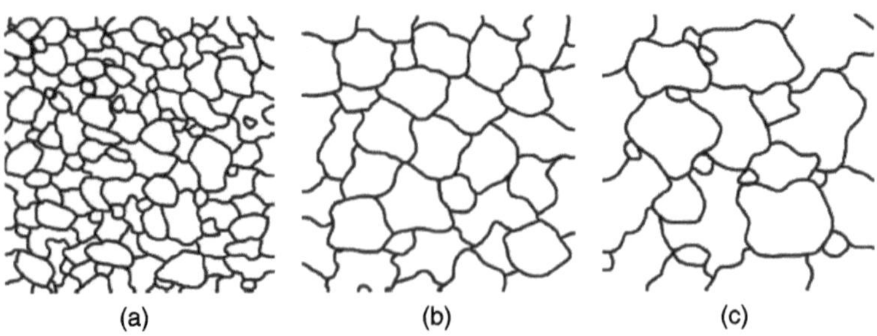

Source: Reprinted from Fátay et al. (2007) [22] with permission from Elsevier.

maximum capacity. For example, an MgH_2 inclusion inside a crystallite, with a radius of 10% of the initial crystallite radius, has only a 0.1% contribution to the total volume of the crystallite. These inclusions appear as very small crystallites in the microstructure after hydrogenation in the second cycle (see Figure 8.6c), therefore yielding the broadening of crystallite size distribution. Contrary to the variation of the crystallite size distribution inside the particles with increasing number of cycles, the size and the morphology of powder particles remain unchanged during repeated dehydriding/hydriding. It is noted that despite the smaller crystallite size in the as-milled sample, the desorption lasts longer in the first cycle than during the subsequent cycles because of the non-activated surface of particles in the beginning of the first dehydrogenation. It seems that the addition of catalyst (e.g. Nb_2O_5 to MgH_2) hinders the growth of crystallites during absorption–desorption cycles at high temperatures [23].

The variation of the crystallite size in the ball-milled MgH_2 within one full desorption–absorption cycle is investigated during the fourth cycle. Figures 8.7a and 8.7b show the variation of the crystallite size of MgH_2 determined during desorption and absorption, respectively. It can be seen that the initial value of 20 nm remains practically unchanged up to 40% of the transformed volume fraction of Mg. At the fraction of 80% Mg, the remaining small amount of MgH_2 has a very small crystallite size of 3 nm [24]. A considerably different crystallite size evolution takes place during the transformation of Mg into MgH_2 (absorption), as it increases almost linearly up to 19 nm when the amount of MgH_2 reaches 100%. The change of the crystallite size during desorption can be explained by the CV theory of transformation of MgH_2 into Mg. If the transformation of MgH_2 to Mg spreads from the boundary into the interior of a crystallite according to the CV model, the volume fraction of the remaining MgH_2 phase ($V_{hydride}$) decreases as:

$$V_{hydride} = \left(\frac{d}{d_0}\right)^3, \quad (8.6)$$

where d_0 and d are the initial (before desorption) and the actual crystallite sizes of the hydride phase, respectively. The normalised crystallite size, $\frac{d}{d_0}$ can be expressed as a function of the fraction of the formed Mg phase ($V_{Mg} = \Delta - V_{hydride}$) by the following formula:

$$\frac{d}{d_0} = (1 - V_{Mg})^{1/3}. \quad (8.7)$$

Figure 8.8 shows the normalised crystallite size as a function of V_{Mg} according to Equation 8.7. Similar to the experimental results presented in Figure 8.7a, the crystallite size decreases slowly in the beginning of desorption while it falls steeply to a very small value at the end of transformation. Considering

Figure 8.7 **Variation of the average crystallite size of MgH_2 as a function of the fraction of Mg and MgH_2 during (a) desorption and (b) absorption, respectively**

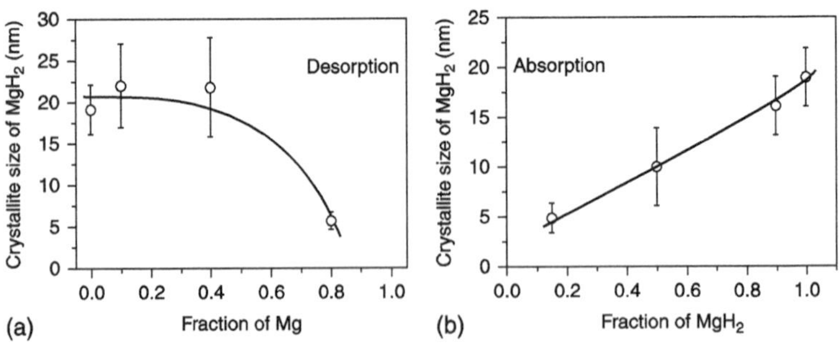

Source: Reprinted from Révészt and Fátay (2010) [24] with permission from Elsevier.

the size distribution of crystallites in MgH_2 phase, the smallest crystallites in the distribution disappear first during dehydrogenation, thereby resulting in a larger average crystallite size of MgH_2 for low transformed fractions of Mg than the values presented in Figure 8.8. This effect lowers the rate of the decrease of crystallite size in the beginning of transformation, yielding an apparent plateau as it was found experimentally (see Figure 8.7a). The dislocation density in the tetragonal MgH_2 phase was not determined but in Mg after dehydrogenation the dislocation density was as small as 6×10^{13} m^{-2}.

In addition to the crystallite size, the type and densities of lattice defects (dislocations, stacking faults and twin boundaries) may change during hydrogenation and dehydrogenation processes. For instance, <a>-type and <c>-type edge dislocations with the Burgers vectors of $\frac{1}{3}\left\langle 11\bar{2}0 \right\rangle$ and $\langle 0001 \rangle$ on the $\left\{01\bar{1}0\right\}$ prismatic slip plane are

Figure 8.8 **Variation of the normalised crystallite size of MgH_2 as a function of the transformed fraction of Mg during desorption according to Equation 8.7**

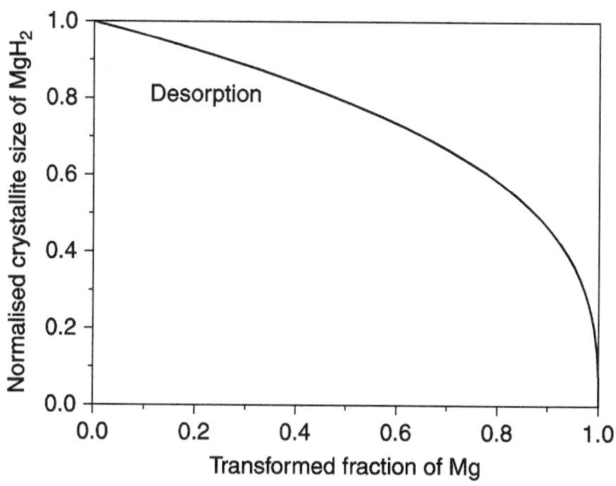

formed in coarse-grained hexagonal $LaNi_5$, $LaNi_{4.5}Cu_{0.5}$ and $LaNi_{4.5}Fe_{0.5}$ during the first absorption process [25–27]. It is revealed that the density of these dislocations is ~10^{15} m^{-2} and they are introduced in order to accommodate the lattice mismatch between the metal–hydrogen solid solution and the metal hydride. It is proposed that the formation of these misfit dislocations needs an excess pressure, which is the main reason for the much higher pressure during the first absorption compared with the first desorption (see the hysteresis in Figure 8.1). In contrast, dislocations are not observed in $LaNi_{4.5}Si_{0.5}$, $LaNi_{4.5}Al_{0.5}$ and $LaNi_{4.75}Sn_{0.25}$, which show little or no hysteresis in the hydrogen pressure-composition isotherms [27]. In these alloys, misfit strains are accommodated by crack formation that occurs more readily than the multiplication of dislocations during hydrogenation. During hydrogen absorption–desorption in coarse-grained Ti-V-Mn alloy with body-centred cubic (bcc) structure, stacking faults and twin boundaries are also introduced in addition to dislocations. These twin boundaries and stacking faults are formed on the {110} and {111} planes of the bcc alloy and the face-centred cubic (fcc) hydride, respectively, in order to accommodate shear strains developed during phase transformation [28].

The defect structure evolved in the hydride phase during the absorption process is strongly influenced by the microstructure of the initial material. This effect was investigated carefully for Mg_2Ni [29]. Mg_2Ni is also a frequently used hydrogen storage material as its hydride is less stable than that of Mg, thereby accelerating the desorption and/or lowering the dehydrogenation temperature, although the storage capacity is smaller. For instance, Mg_2Ni is able to form Mg_2NiH_4 with a hydrogen storage capacity of 3.6 wt%. The hydrogen release of this hydride occurs at a lower temperature (about 250°C) than that of MgH_2 [29]. When ball-milled Mg_2Ni with 15–20 nm crystallite size is hydrogenated at 200°C, monoclinic low-temperature phase of Mg_2NiH_4 forms which contains a much larger amount of stacking faults and/or twin boundaries than the hydride obtained by hydrogenation of coarse-grained Mg_2Ni at the same temperature. Most probably, the large distortions in the ball-milled Mg_2Ni are relaxing by the formation of stacking faults and twins during hydrogenation. Dislocations, stacking faults and twin boundaries with very large densities may also form in hydrogen storage materials processed without application of severe plastic deformation. For example, after dehydrogenation of Mg_2NiH_4 produced by hydriding combustion synthesis, Mg_2Ni is obtained with very high dislocation density of $\sim 10^{17}$ m^{-2}, which develops due to stresses induced by phase transformations in combustion synthesis and subsequent dehydrogenation [30]. The high defect density improves the sorption kinetics by increasing diffusivity and the number of potential sites for new phase nucleation.

8.4 Effect of defects on hydrogen storage properties of carbon nanotubes

Carbon nanotubes (CNTs) are obtained in two different main forms characterised by the structure of their wall: single-walled nanotubes (SWNTs) and multi-walled nanotubes (MWNTs). SWNTs consist of a graphene sheet rolled up into a cylinder of a few nanometres' diameter and several microns' length [31]. A MWNT is an arrangement of coaxial SWNTs ranging in number from 2 up to about 50. The diameters of these MWNT range from a few to tens of nanometres and their length is of the order of a micrometre.

There are many methods to produce CNTs among which CNTs grown by chemical vapour deposition (CVD) have higher defect density compared with CNTs synthesised by arc discharge and laser ablation methods [32]. Defects can also be introduced after growth of CNTs by different methods such as milling [33], oxidation [34], nitric acid treatment [31], potassium hydroxide treatment [35], plasma etching [36] and irradiation [37]. During heat-treatment of pristine CNTs, both defect generation and defect annealing processes take place [32].

Due to the lightweight and high specific surface area of carbon nanotubes (CNTs), they are good candidates for hydrogen storage. In SWNTs, hydrogen can be adsorbed at the exterior of the tube wall by forming H-C bonds or in the middle of the tubes by H-H bonds. The adsorption into the interior wall of the tube is also possible but not stable and hydrogen form H-H bonds inside the tube [38]. MWNTs can adsorb hydrogen between the walls of the coaxial tubes. The amount of hydrogen stored by both physisorption due to van der Waals forces and chemisorption in as-prepared CNTs is usually less than 1 wt.%, although in some cases higher storage capacities of 4–7 wt.% are reported [38]. The scattering in the experimentally obtained capacity values can be explained by the non-uniform morphology of CNTs (diameter, length and wall thickness, etc.) and the variation of the purity level [34]. Opening of CNTs by removing buckyball structured caps from the ends of nanotubes by milling, nitric acid treatment or oxidative technique improves storage capacity because hydrogen molecules cannot penetrate through the carbon atom hexagons in the wall [31]. Additionally, the introduction of defects into carbon nanotubes activates hydrogen adsorption and enhances the chemisorption, thereby yielding a large enhancement of hydrogen adsorption and desorption. Defects in CNTs can be produced by ion irradiation at keV to MeV energies [37]. High energy ions produce atomic displacements within the six-membered carbon rings. In addition, the displaced atoms or knock-ons may have sufficient energy to displace additional atoms. The damaged region is collectively referred to as a collision cascade and consists of single vacancies, divacancies and displaced atoms. The knock-on atoms will lose their energy through multiple scattering until they are thermalised and adsorbed onto the nanotube walls as adatoms. These defects in SWNTs are illustrated in Figure 8.9a. Subsequent thermal treatment after ion irradiation may transform the vacancies and the adatoms into non-hexagonal rings and/or pentagon/heptagon Stone-Wales (SW) defects as illustrated in Figure 8.9b. For comparison of the hydrogen storage capacity of SWNTs irradiated with a fluence of 5×10^{16} H-ions/cm^2 with a non-treated sample, first both specimens

Figure 8.9 **Front walls of one and the same SWNT just after ion impact (a) and after annealing (b). During annealing the divacancy (D) transformed to an agglomeration of non-hexagonal rings. The numbers in the centre of the rings indicate the numbers of atoms constituting the rings. The single vacancy (S) and the nearby carbon adatom (A) in the upper right-hand corner in (a) transformed to a Stone-Wales 5–7 defect in (b)**

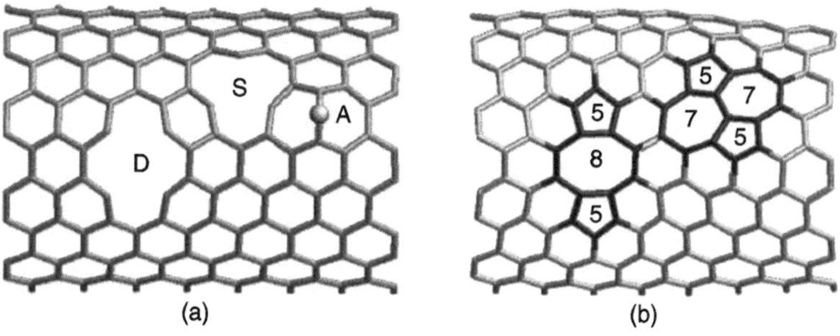

Source: Reprinted from Krasheninnikov and Nordlund (2004) [39] with permission from Elsevier.

were hydrogenated under a pressure of 83 bar and at 200°C for 12 hours. Then, during desorption at 450°C 8 times larger amount of hydrogen released from the irradiated material than from the untreated CNTs.

Pores and defects on MWNT walls can also be generated by pulse oxidation at 600°C, as seen in Figure 8.10a [34]. The pristine MWNTs can uptake only 0.1 wt.% hydrogen at 300°C under ambient pressure while the defective MWNTs can store 0.5 wt.% hydrogen, indicating that to create defects on MWNTs is an effective route for enhancing the hydrogen storage. The capacity of the pristine MWNTs can be increased from 0.1 wt.% to 1.8 wt.% by doping with Pd–Ni catalyst, which promotes adsorption and dissociation of hydrogen molecules. When the defective MWNTs are doped with Pd–Ni, the superposition of the effects of the defects and the catalyst yields a very high capacity of 6.6 wt.% hydrogen. Pd–Ni doping over defective MWCNTs not only increases the hydrogen storage capacity, but also decreases the hydrogen release temperature from 677°C to 337°C. Pd–Ni alloy forms nanoparticles with the size of 10–20 nm on the walls of MWNTs having similar diameters (see Figure 8.10b). The majority of hydrogen is stored in the form of CH and CH_2 species attached to the walls of nanotubes. The carbon atoms located at the defective parts of CNTs have a dangling bond and hence are able to associate with atomic hydrogen to form CH or CH_2 species [34]. This chemisorption is the main reason for the increased capacity of defective CNTs with catalyst. Without defects and the metal catalyst assistance, the hydrogen molecule is not easily dissociated to atomic hydrogen over the CNTs and unlikely to form the C–H bond. As a consequence, in pristine CNTs hydrogen is rather stored in molecular form with the help of physisorption.

Pores on the outer surface of MWNTs facilitate the hydrogen spillover process that comprises the dissociation of

Figure 8.10 **TEM images of (a) defective MWCNTs processed by oxidation, (b) defective MWCNTs doped with Pd–Ni catalyst nanoparticles (appearing as black dots)**

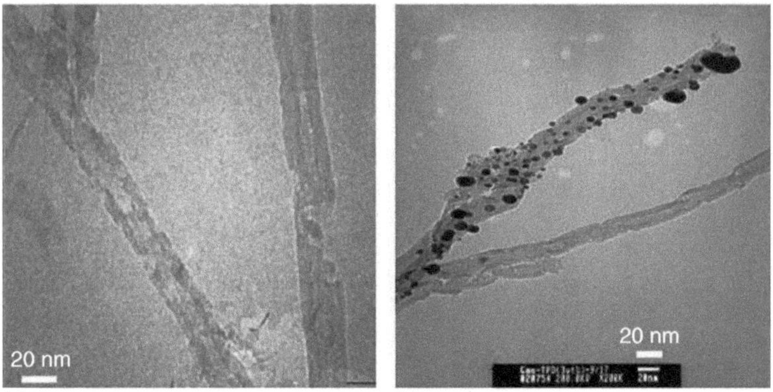

(a) (b)

Source: Reprinted from Gao et al. (2010) [34] with permission from Elsevier.

hydrogen molecules to atomic hydrogen on a metal catalyst and subsequent migration from the metal to the surface of CNTs [35]. Figure 8.11 schematically compares hydrogen spillover processes for defect-free and defective MWNTs. In the former case, the dissociated hydrogen atoms can migrate from the catalyst particles only to the external surface of the CNTs. At the same time, for a MWNT containing pores in the vicinity of catalyst particles, hydrogen atoms easily diffuse and migrate from the external surface into interplanar spacing via the pores, causing hydrogen spillover enhancement and therefore larger storage capacity.

Ball-milling provides a simple and efficient method to introduce defects into CNTs. The collision and friction with the milling balls result in: (i) shortening of CNTs by cleavage leading to an increase of specific surface area; (ii) opening the ends of CNTs; and (iii) increase of defect density. As a consequence, milling improves the hydrogen adsorption properties of CNTs. For instance, the hydrogen adsorption

Figure 8.11 Schematic representation of hydrogen spillover in (a) defect-free and (b) defective MWNTs decorated by catalyst particles. H: atomic hydrogen, H_2: hydrogen molecule

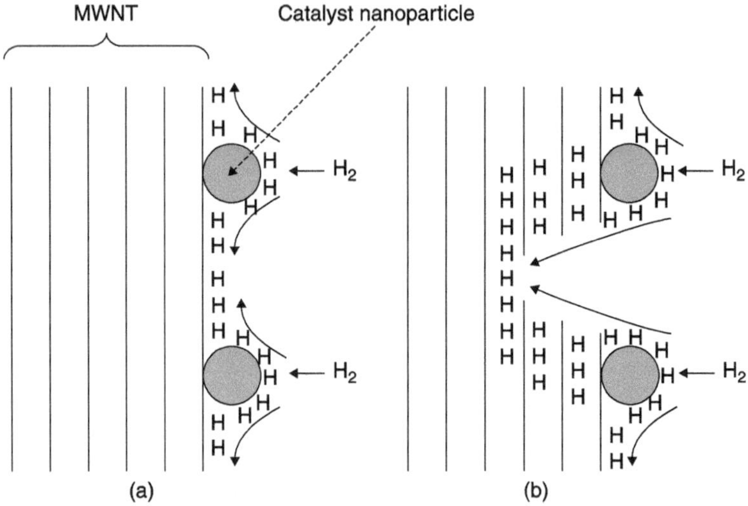

Source: Reprinted from Chen and Huang (2008) [35] with permission from Elsevier.

capacity of MWNTs at room temperature under a pressure of 8–9 MPa increases from 0.1 wt.% to 0.66 wt% due to milling for 10 hours [33]. The variation of capacity is non-monotonous as first it increases with milling time up to ~10 hours, then decreases to ~0.2 wt.% after 30–40 hours' milling and remains unchanged for longer milling times. The decrease of storage capacity for long milling times can be explained by the formation of amorphous carbon having poor hydrogen adsorption properties [33].

Defects on the walls of CNTs can also be introduced by plasma etching. In this process the plasma can be produced in hydrogen gas using a microwave plasma generator [36]. The non-etched MWNTs have ~0.6 wt.% hydrogen uptake capacity at ambient temperature and under the pressure of

~11 MPa, while after 2 hours' and 4 hours' etching processes, the hydrogen storage capacity reaches ~1.0 wt.% and ~1.4 wt.%, respectively. In addition to the promotion effect of defects on the chemisorption of hydrogen, they also facilitate the access of hydrogen molecules to interlayers and hollow interiors of MWNTs. Without defects on CNT walls, hydrogen has to enter the tubes through the open ends. Additionally, the pores on the external surface of MWNTs can store hydrogen molecules and their contribution to hydrogen capacity linearly increases with the volume fraction of pores [40]. The decoration of MWNTs by Pd nanoparticles after 2 hours' and 4 hours' etching yields an additional increase of hydrogen storage capacity to 4 wt.% and 4.5 wt.%, respectively [36]. Pd catalyst promotes the dissociation of H_2 into hydrogen atoms that easily penetrate tube walls and are combined into H_2 in interlayers and hollow interiors of CNTs or form C-H bonds by chemisorption.

All-electron density functional theory calculations predict good ability of hydrogen storage for defective nitrogen-doped CNTs decorated by Ca atoms [41]. Nitrogen atoms substitute carbon atoms in divacancies and Ca-adsorption on the hollow site of these defects is energetically favourable. According to the calculations, in a H_2 adsorption process the optimised structure is reached when five hydrogen molecules are attached to one Ca atom at divacancies, which yields a theoretical storage capacity of ~3 wt.% without any catalyst.

8.5 References

1. R.A. Varin, T. Czujko and Z.S. Wronski, *Nanomaterials for Solid State Hydrogen Storage*, Springer Science+Business Media, LLC 2009, ISBN 978-0-387-77711-5.

2. M. Dornheim, S. Doppiu, G. Barkhordarian, U. Boesenberg, T. Klassen, O. Gutfleisch and R. Bormann, 'Hydrogen storage in magnesium-based hydride composites', *Scripta Materialia*, 56 (2007) 841–846.
3. E. David and J. Mater, 'An overview of advanced materials for hydrogen storage', *Journal of Materials Processing Technology*, 162–163 (2005) 169–177.
4. G. Barkhordarian, T. Klassen and R. Bormann, 'Kinetic investigation of the effect of milling time on the hydrogen sorption of magnesium catalyzed with different Nb2O5 contents', *Journal of Alloys and Compounds*, 407 (2006) 249–255.
5. J.L. Bobet, C. Even, Y. Nakamura, E. Akiba and B. Darriet, 'Synthesis of magnesium and titanium hydride via reactive mechanical alloying. Influence of 3d-metal addition on MgH2 synthesize', *Journal of Alloys and Compounds*, 298 (2000) 279–284.
6. G. Barkhordarian, T. Klassen and R. Bormann, 'Effect of Nb205 content on hydrogen reaction kinetics of Mg', *Journal of Alloys and Compounds*, 364 (2004) 242–246.
7. D. Fatay, A. Revesz and T. Spassov, 'Particle size and catalytic effect on the dehydriding of MgH2', *Journal of Alloys and Compounds*, 399 (2005) 237–241.
8. J. Topler, H. Buchner, H. Saufferer, K. Knorr and W. Prandl, 'Measurement of the diffusion of hydrogen atoms in magnesium and Mg2Ni by neutron scattering', *Journal of The Less-Common Metals*, 88 (1982) 397–404
9. Á. Révész, D. Fátay and T. Spassov, 'Hydriding kinetics of ball-milled nanocrystalline MgH2 powders', *Journal of Materials Research*, 22 (2007) 3144–3151.
10. V.M. Skripnyuk, E. Rabkin, Y. Estrin and R. Lapovok, 'The effect of ball milling and equal channel angular pressing on the hydrogen absorption/desorption

properties of Mg-4.95 wt% zn-0.71 wt% Zr (ZK60) alloy', *Acta Materialia*, 52 (2004) 405–414.

11. V. Skripnyuk, E. Buchman, E. Rabkin, Y. Estrin, M. Popov and S. Jorgensen, 'The effect of equal channel angular pressing on hydrogen storage properties of eutectic Mg-Ni alloy', *Journal of Alloys and Compounds*, 436 (2007) 99–106.
12. Y. Kusadome, K. Ikeda, Y. Nakamori, S. Orimob and Z. Horita, 'Hydrogen storage capability of MgNi2 processed by high pressure torsion', *Scripta Materialia*, 57 (2007) 751–753.
13. V.M. Skripnyuk, E. Rabkin, Y. Estrin and R. Lapovok, 'Improving hydrogen storage properties of magnesium based alloys by equal channel angular pressing', *International Journal of Hydrogen Energy*, 34 (2009) 6320–6324.
14. M. Zehetbauer, R. Grossinger, H. Krenn, M. Krystian, R. Pippan, P. Rogl, T. Waitz and R. Wurschum, 'Bulk Nanostructured Functional Materials by Severe Plastic Deformation', *Advanced Engineering Materials*, 12 (2010) 692–700.
15. Á. Révész, Zs. Kánya, T. Verebélyi, P.J. Szabó, A.P. Zhilyaev and T. Spassov, 'The effect of high-pressure torsion on the microstructure and hydrogen absorption kinetics of ball-milled Mg70Ni30', *Journal of Alloys and Compounds*, 504 (2010) 83–88.
16. K. Edalati, A. Yamamoto, Z. Horita and T. Ishihara, 'High-pressure torsion of pure magnesium: Evolution of mechanical properties, microstructures and hydrogen storage capacity with equivalent strain', *Scripta Materialia*, 64 (2011) 880–883.
17. M. Danaie, C. Mauer, D. Mitlin and J. Huot, 'Hydrogen storage in bulk Mg-Ti and Mg-stainless steel multilayer composites synthesized via accumulative roll-bonding

(ARB)', *International Journal of Hydrogen Energy*, 36 (2011) 3022–3036.

18. C. Delmas and C. Tessier, 'Stacking faults in the structure of nickel hydroxide: a rationale of its high electrochemical activity', *Journal of Materials Chemistry*, 7 (8) (1997) 1439–1443.
19. G.J.C. Carpenter and Z.S. Wronski, 'Nanocrystalline NiO and NiO-Ni(OH)2 composite powders prepared by thermal and mechanical dehydroxylation of nickel hydroxide', *Nanostructured Materials*, 11 (1) (1999) 67–80.
20. F. Bernard, E. Gaffet, Y. Champion, N. Boudarina and A. Ustinov, 'Correlation between ball milling conditions and planar effects on Cu-nanostructured powders', *Journal of Physics*, IV 12 (2002) 455–460.
21. P.A. Huhn, M. Dornheim, T. Klassen and R. Bormann, 'Thermal stability of nanocrystalline magnesium for hydrogen storage', *Journal of Alloys and Compounds*, 404–406 (2005) 499–502.
22. D. Fátay, T. Spassov, P. Delchev, G. Ribárik and Á. Révész, 'Microstructural development in nanocrystalline MgH2 during H-absorbtion/desorbtion cycling', *International Journal of Hydrogen Energy*, 32 (2007) 2914–2919.
23. O. Friedrichs, F. Aguey-Zinsou, J.R. Ares Fernandez, J.C. Sanchez-Lopez, A. Justo, T. Klassen, R. Bormann and A. Fernandez, 'MgH2 with Nb2O5 as additive, for hydrogen storage: Chemical, structural and kinetic behaviour with heating', *Acta Materialia*, 54 (2006) 105–110.
24. Á. Révész and D. Fátay, 'Microstructural evolution of ball-milled MgH2 during a complete dehydration-hydrogenation cycle', *Journal of Power Sources*, 195 (2010) 6997–7002.
25. H. Inui, T. Yamamoto, M. Hirota and M. Yamaguchi, 'Lattice defects introduced during hydrogen absorption-

desorption cycles and their effects on P-C characteristics in some intermetallic compounds', *Journal of Alloys and Compounds*, 330–332 (2002) 117–124.
26. B. Decamps, J.-M. Joubert, R. Cerny and A. Percheron-Guegan, 'TEM study of the dislocations generated by hydrogen absorption/desorption in LaNi5 and derivatives', *Journal of Alloys and Compounds*, 404–406 (2005) 570–575.
27. J. Matsuda, Y. Nakamura and E. Akiba, 'Lattice defects introduced into LaNi5-based alloys during hydrogen absorption/desorption cycling', *Journal of Alloys and Compounds*, 509 (2011) 7498–7503.
28. J. Matsuda, Y. Nakamura and E. Akiba, 'Microstructure of Ti-V-Mn BCC alloys before and after hydrogen absorbtion-desorption', *Journal of Alloys and Compounds*, 509 (2011) 4352–4356.
29. R. Janot, F. Cuevas, M. Latroche and A. Percheron-Guegan, 'Influence of crystallinity on the structural and hydrogenation properties of Mg2X phases (X=Ni, Si, Ge, Sn)', *Intermetallics*, 14 (2006) 163–169.
30. D. Liu, Y. Zhu and L. Li, 'Crystal defect analysis and surface characteristics of Mg_2NiH_4 produced by hybriding combustion synthesis', *International Journal of Hydrogen Energy*, 32 (2007) 2417–2421.
31. F. Lamari Darkrim, P. Malbrunot and G.P. Tartaglia, 'Review of hydrogen storage by absorbtion in carbon nanotubes', *International Journal of Hydrogen Energy*, 27 (2002) 193–202.
32. G.M. Bhalerao, A.K. Sinha and V. Sathe, 'Defect-dependent annealing behaviour of multi-walled carbon nanotubes', *Physics E: Low-dimensional Systems and Nanostructures*, 41 (2008) 54– 59.
33. F. Liu, X. Zhang, J. Cheng, J. Tu, F. Kong, W. Huang and C. Chen, 'Preparation of short carbon nanotubes by

mechanical ball milling and their hydrogen absorbtion behaviour', *Carbon*, 41 (2003) 2527–2532.

34. L. Gao, E. Yoo, J. Nakamura, W. Zhang and H.T. Chua, 'Hydrogen storage in Pd-Ni doped defective carbon nanotubes through the formation of CHx (x=1,2)', *Carbon*, 48 (2010) 3250–3255.
35. C.-H. Chen and C.-C. Huang, 'Enhancement of hydrogen spillover onto carbon nanotubes with defect feature', *Microporous and Mesoporous Materials*, 109 (2008) 549–559.
36. S-C. Mu, H-L. Tang, S-H. Qian, M. Pan and R-Z. Yuan, 'Hydrogen storage in carbon nanotubes modified by microwave plasma etching and Pd decoration', *Carbon*, 44 (2006) 762–767.
37. F.D. McDaniel, F.U. Naab, O.W. Holland, M. Dhoubhadel, L.J. Mitchell and J.L. Duggan, 'Low-energy ion irradiation effects on hydrogen absorption and desorption in carbon nanotubes', *Surface and Coatings Technology*, 201 (2007) 8564–8567.
38. E. David, 'An overview of advanced materials for hydrogen storage', *Journal of Materials Processing Technology*, (2005) 169–177.
39. A.V. Krasheninnikov and K. Nordlund, 'Irradiation effects in carbon nanotubes', *Nuclear Instruments and Methods in Physics Research, Section B: Beam Interactions with Materials and Atoms*, 216 (2004) 355–366.
40. C.-H. Chen and C.-C. Huang, 'Hydrogen absorption in defective carbon nanotubes', *Separation and Purification Technology*, 65 (2009) 305–310.
41. J.-X. Zhao, Yi-H. Ding, X.-G. Wang, Q.-H. Cai and X.-Z. Wang, 'Theoretical studies of the CNx nanotube with four nitrogen divacancy (4ND) defects', *Diamond and Related Materials*, 20 (2011) 36–41.

Appendix: characterisation of defect structure by x-ray diffraction line profile analysis

Abstract: X-ray diffraction line profile analysis is a powerful method for the characterisation of defect structure of nanomaterials. The most important advantages of this method are: (i) the easy sample preparation; (ii) its non-destructivity; and (iii) the good statistics of the resulted microstructural parameters compared with microscopic methods. In this appendix, the principle of x-ray line profile analysis is presented since the majority of the results presented in this book are obtained by this technique. The parameters of the defect structure determinable by x-ray line profile analysis are described in detail.

Key words: x-ray line profile analysis; crystallite size distribution; dislocation density; edge/screw type of dislocation; dislocation arrangement parameter; twin boundary frequency.

The lattice defect structure in ultrafine-grained (UFG) and nanocrystalline materials can be investigated by direct methods such as transmission electron microscopy (TEM) or by indirect techniques, e.g. by x-ray diffraction line profile analysis (XLPA). Using the microscopy methods, a direct image of the microstructure can be observed, while in XLPA first only a fingerprint of the microstructure – a diffraction pattern – is obtained and then the parameters of the defect structure are extracted by analysing the breadth and the shape of the diffraction peak profiles. In UFG and nanocrystalline materials x-ray diffraction peak profiles broaden mainly due to the (i) small crystallite size, (ii) dislocations and (iii) planar defects, especially stacking faults and twin boundaries. Additionally there may be other sources of line broadening such as chemical heterogeneities [1] and/or surface relaxation in nanoparticles [2], but these effects were neglected beside the broadening caused by the lattice defects in the materials studied in this book. The parameters of the microstructure in XLPA are usually determined by fitting theoretical functions on the measured peak profiles. These functions are calculated on the basis of a model of the microstructure. For the construction and control of realistic models, simultaneous TEM investigations are suggested. The most important advantages of XLPA are its non-destructivity and the very good statistics of the parameters characterising the defect structure, since x-ray diffraction usually examines 10^3–10^5 larger sample volume than in TEM [3, 4]. It should be noted that XLPA can provide the dislocation density if its value is larger than ~10^{13} m^{-2}, since under this limit the peak broadening caused by dislocations is masked by the instrumental breadth. At the same time, conventional TEM can determine a reliable average dislocation density if its value is smaller than ~10^{14} m^{-2} because above this limit the contrasts of individual dislocations strongly overlap. The instrumental effect in XLPA also yields upper and lower detection limits of the crystallite size and the twin boundary frequency, respectively. These detection limits strongly depend on the instrument but their values are usually ~400–800 nm and ~0.02–0.1% for the mean crystallite size and the twin boundary frequency, respectively. It is noted that the crystallite size measured by

XLPA is usually smaller than the grain size obtained by TEM for bulk UFG metals processed by severe plastic deformation methods (e.g. equal-channel angular pressing). This phenomenon can be attributed to the fact that the crystallites are equivalent to the domains in the microstructure which scatter x-rays coherently. As the coherency of x-rays breaks even if they are scattered from volumes having quite small misorientations (1–2°), the crystallite size corresponds rather to the subgrain size in the severely deformed microstructures [4]. At the same time, for UFG or nanocrystalline materials processed by condensation or deposition methods, the crystallite size obtained by XLPA agrees well with the grain size determined by TEM due to the lack of subgrain structure. For milled nanopowders, the crystallite and grain sizes are in good agreement, although they are about two orders of magnitude smaller than the particle size.

According to the kinematical theory of x-ray scattering, the intensity profile of a diffraction peak is the convolution of the instrumental profile, the size profile caused by the small crystallite size, the distortion or strain profile resulted from the strain field of dislocations and the profile function caused by planar faults (stacking faults and/or twin boundaries). Consequently, the Fourier-transform of the measured intensity profile, A, is the product of the Fourier-transforms of the instrumental (A^I), size (A^S), distortion (A^D) and planar fault (A^F) profiles [3, 4]:

$$A(L) = A^I(L)A^S(L)\,A^D(L)\,A^F(L), \qquad \text{(A1)}$$

where L is the Fourier variable with length dimension. The instrumental broadening for a given equipment can be measured on a standard material which has large crystals with a small amount of lattice defects. The measurement for the standard material should be run under the same instrumental conditions as for the specimen studied, which ensure that the broadening of the standard exactly gives the instrumental broadening in the x-ray diffraction pattern of the specimen. For the determination of the instrumental broadening the most often used material is the LaB_6 standard (NBS 660a). An extra advantage of the usage of LaB_6 as a profile standard material is that it has many peaks appearing more or less equidistantly in the diffraction pattern.

According to the theorem of Bertaut [5, 6], the size profile of the diffraction peak with the indices *hkl* is the same as the sum of the intensities diffracted incoherently by the columns building up the reflecting crystallites and having infinitesimal cross-sectional area as well as aligned normal to the reflecting (*hkl*) lattice planes (i.e. parallel to the *hkl* diffraction vector **g**). The division of the reflecting crystallites into scattering columns according to the Bertaut-theorem is illustrated in Figure A1. As a consequence, the Fourier transform of the size profile can be calculated as the sum of the Fourier transforms of the intensities scattered by the individual

Figure A1 The division of reflecting crystallites into scattering columns according to the Bertaut theorem

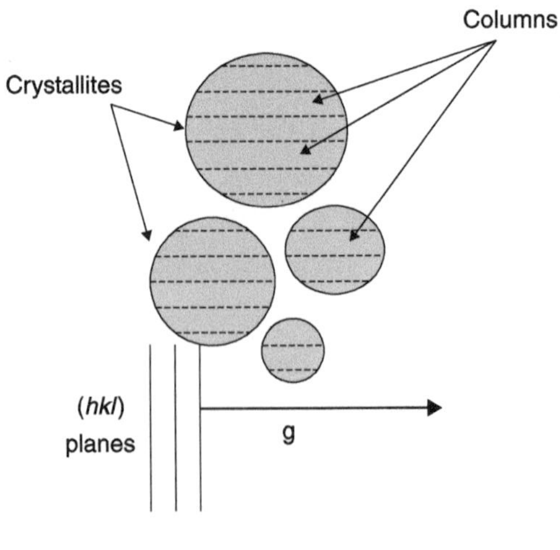

columns. Using the column length distribution density function *p(t)*, the Fourier transform of the *hkl* size profile is given as [6]:

$$A^S(L) = \frac{1}{\langle t \rangle}\int_L^{\infty}(t-L)p(t)\,dt, \tag{A2}$$

where *p(t)dt* represents the relative fraction of columns for which the length parallel to the *hkl* diffraction vector lies between *t* and *t+dt* while *<t>* is an average column length obtained as the ratio of the total volume of the reflecting crystallites and their cross-section normal to the *hkl* diffraction vector. In practice, for a given shape and size distribution of crystallites the function *p(t)* is expressed by the crystallite size distribution density function, *f(x)*, as [6]:

$$p(t) = N\int_0^{\infty} h(t,x)f(x)\,dx, \tag{A3}$$

where *N* is a normalisation factor and *h(t,x)dt* is the relative fraction of columns having lengths between *t* and *t+dt* in the crystallite with diameter *x* [7]. For spherical crystallites with log-normal size distribution:

$$h(t,x) = \frac{2t}{x^2} \tag{A4}$$

for $t \le x$ and $h(t,x) = 0$ for $t > x$, and:

$$f(x) = \frac{1}{\sqrt{2\pi}\sigma x}\exp\left\{-\frac{[\ln(x/m)]^2}{2\sigma^2}\right\}. \tag{A5}$$

where *m* and σ are the median and the variance of the log-normal crystallite size distribution, respectively. Substituting Equations A4 and A5 into Equation A3 and then Equation A3 into Equation A2, the size Fourier transform is written as [8, 9]:

$$A^S(L) = \frac{1}{2} erfc\left[\frac{\ln(|L|/m)}{\sqrt{2}\sigma} - 1.5\sqrt{2}\sigma\right] -$$

$$-\frac{3}{4m\exp(8.125\sigma^2)} |L| \, erfc\left[\frac{\ln(|L|/m)}{\sqrt{2}\sigma} - \sqrt{2}\sigma\right] + \qquad \text{(A6)}$$

$$+\frac{3}{4m^3\exp(10.125\sigma^2)} |L|^3 erfc\left[\frac{\ln(|L|/m)}{\sqrt{2}\sigma}\right],$$

where *erfc* is the complementary error function given as:

$$erfc(x) = \frac{2}{\sqrt{\pi}} \int_x^{\infty} \exp(-t^2)\, dt. \qquad \text{(A7)}$$

Thus, both the size Fourier transform, $A^S(L)$ and the size peak profile that can be obtained as the inverse Fourier transform of $A^S(L)$ depend only on the two parameters (m and σ) of the size distribution of spherical crystallites. It is emphasised that the peak broadening of the size profile with *hkl* indices is determined by the size of the reflecting crystallites measured parallel to the *hkl* diffraction vector, i.e. normal to the (*hkl*) lattice planes. The smaller the crystallite size, the broader the diffraction peak.

The Fourier transform of the strain or distortion peak profile can be approximated by the following formula [10]:

$$A^D(g,L) = \exp\left(-2\pi^2 L^2 g^2 \left\langle \varepsilon_{g,L}^2 \right\rangle\right), \qquad \text{(A8)}$$

where g is the length of the *hkl* diffraction vector $\left\langle g = \frac{2\sin\Theta}{\lambda}\right\rangle$ where Θ is the Bragg-angle of the diffraction peak and λ is the wavelength of x-rays) and $\left\langle \varepsilon_{g,L}^2 \right\rangle$ is the mean-square strain normal to (*hkl*) lattice planes that depends on both **g** and L. For understanding the meaning of $\left\langle \varepsilon_{g,L}^2 \right\rangle$, first let us imagine a distortion-free crystal lattice where the section between points A and B is lying normal to the (*hkl*) lattice planes and has a length of L (see Figure A2). Due to the lattice distortions caused by defects (e.g. dislocations) in the real crystal, the positions of points A and B are shifted to A' and B'. The projection of the section A'B' normal to (*hkl*) lattice planes is L'. The ratio $(L'-L)/L$ gives the local strain normal to (*hkl*) planes. Averaging the square of this quantity for the reflecting volume yields the value of $\langle \varepsilon^2_{g,L} \rangle$. It is emphasised that the strain broadening of *hkl* peak profile is determined by the lattice distortions normal to (*hkl*) lattice planes in the reflecting crystallites. The larger the lattice distortion, the broader the diffraction peak.

The lattice distortions are mainly caused by the strain fields of lattice defects. The strain field of a point defect decreases with r^{-3}, where r is the distance from the defect, therefore it

Figure A2 Characterisation of lattice distortions parallel to the diffraction vector **g** (or normal to the (*hkl*) lattice planes)

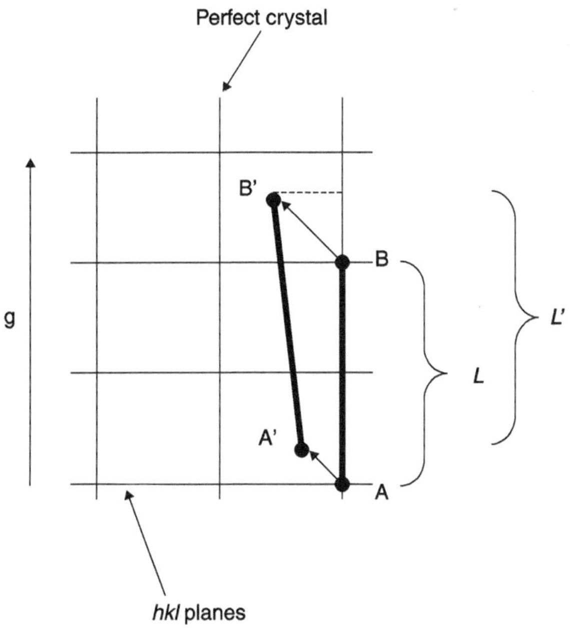

decays very quickly with increasing r. At the same time, the strain field of an individual dislocation is of long-range character as it decays with r^{-1}. Due to the reciprocity between crystal and reciprocal spaces, the scattering related to point defects (referred to as Huang scattering) is extended far from the fundamental Bragg reflections and therefore not involved in the evaluation of line profiles. Due to the long-range character of strain field of dislocations, these defects yield considerable broadening of line profiles. For dislocations, the mean-square strain can be given as:

$$\left\langle \varepsilon_{g,L}^2 \right\rangle = \frac{b^2}{4\pi}\rho C f^*\left(\frac{L}{R_e}\right), \tag{A9}$$

where ρ and b are the dislocation density and the magnitude of Burgers vector, respectively, C is the contrast factor of dislocations and R_e is the effective outer cut-off radius of dislocations with length dimension. The function $f^*\left(\frac{L}{R_e}\right)$ is given in Wilkens (1970) [11]. The parameter R_e reflects that the strain field of a dislocation structure depends not only on the dislocation density but also on the arrangement of dislocations. If the strain fields of the individual dislocations are screening each other, the total distortion in the lattice is lower, which yields a smaller value of R_e. Strong screening of strain fields of dislocations occurs e.g. when the dislocations are arranged into dipoles or low-angle grain boundaries, as depicted in Figure A3. Instead of R_e, the screening of strain fields of dislocations is rather described by the dimensionless dislocation arrangement parameter M, which can be calculated as:

$$M = R_e\rho^{1/2}. \tag{A10}$$

The stronger the screening of the strain fields of dislocation, the smaller the value of M and the longer the tails in the diffraction profiles.

Figure A3 **Schematic illustration of some arrangements of dislocations yielding to weak or strong screening of the strain fields and the corresponding large or small value of the dislocation arrangement parameter *M***

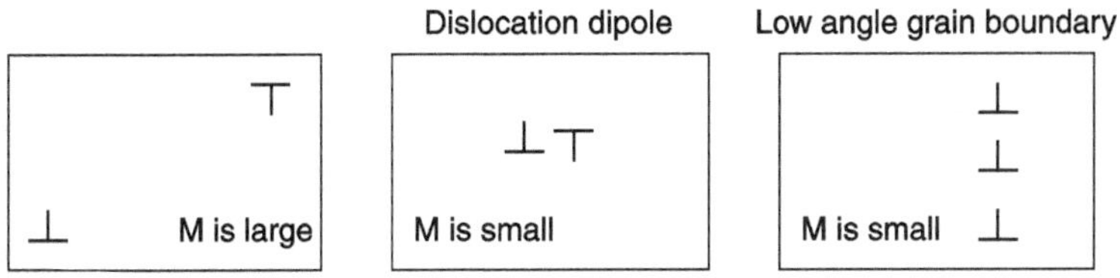

Due to the anisotropic strain field of a dislocation, the mean-square strain is different in the various crystallographic directions for given values of ρ and R_e, which yields a strong dependence of peak broadening on *hkl* indices of reflections. This effect is referred to as strain anisotropy [12, 13] and taken into account by the contrast factor C in Equation A9. For an individual dislocation, the value of C depends on the orientation of the Burgers and line vectors of dislocation, and the diffraction vector. In a texture-free polycrystal or if the different slip systems are equally populated by dislocations, the C factors can be averaged over the permutations of the *hkl* indices [14] and the average contrast factor, $\bar{C}_{hkl}$, should be inserted into Equation A9. For materials with cubic crystal structure, the average contrast factor is given as [14]:

$$\bar{C}_{hkl} = \bar{C}_{h00}(1 - qH^2), \tag{A11}$$

where $\bar{C}_{h00}$ is the average contrast factor for reflection *h00* and:

$$H^2 = \frac{h^2k^2 + h^2l^2 + k^2l^2}{(h^2 + k^2 + l^2)^2}, \tag{A12}$$

where *hkl* are the indices of reflection. The values of parameters $\bar{C}_{h00}$ and q depend on the anisotropic elastic constants of the crystal and the type of prevailing dislocations (e.g. edge or screw). The contrast factors can be determined by the software *Anizc*, which can be accessed on the website *http://metal.elte.hu/anizc* [15]. Assuming $\frac{1}{2}\langle 110\rangle\{111\}$ and $\frac{1}{2}\langle 111\rangle\{110\}$ slip systems for face-centred cubic (fcc) Al and Cu, and body-centred cubic (bcc) Fe, respectively, the values of $\bar{C}_{h00}$ and q for edge and screw dislocations are listed in Table A1.

In hexagonal close-packed (hcp) crystals, the average contrast factor, $\bar{C}_{hkl}$ can be given as [16]:

$$\bar{C}_{hkl} = \bar{C}_{hk0}(1 + q_1 z + q_2 z^2), \tag{A13}$$

where q_1 and q_2 are two parameters depending on the anisotropic elastic constants of the crystal and the type of slip system. $z = (2/3)(l/ga)^2$, where a is the lattice constant in the basal plane. $\bar{C}_{hk0}$ is the average dislocation contrast factor of the *hk0* type reflections. The values of $\bar{C}_{hk0}$, q_1 and q_2 are listed for the 11 most common slip systems in a large number of hexagonal materials in Dragomir and Ungár (2002) [16].

The peak profile caused by planar defects depends on the type of defect (intrinsic or extrinsic stacking fault or twin boundary) and their frequencies. The stacking fault or twin boundary frequency is defined as the relative fraction of the

Table A1 The values of $\bar{C}_{h00}$ and q for edge and screw dislocations in Al, Cu and Fe

	$\bar{C}_{h00}$ (edge / screw)	q (edge / screw)
Al	0.2011 / 0.1839	0.36 / 1.33
Cu	0.3076 / 0.3029	1.67 / 2.33
Fe	0.2548 / 0.3013	1.28 / 2.67

faulted planes along their normal vector and usually it is given as a percentage. The formulas of peak profiles corresponding to planar faults in fcc and hcp crystals are given in Balogh et al. (2006) [17] and Balogh et al. (2009) [18], respectively. In order to compare the planar fault frequency determined by XLPA with the direct observation performed by TEM, the planar fault frequency can be transformed into the mean fault-spacing. For instance, in fcc crystals the mean twin-spacing (d_{twin}) can be obtained from the twin boundary frequency (β) as:

$$d_{twin} = \frac{100 \cdot d_{111}}{\beta} \tag{A14}$$

where d_{111} is the distance between the neighbouring {111} planes. The mean twin-spacing values obtained by XLPA and TEM show good agreement within the experimental error, as illustrated in Figure A4.

Figure A4 The correlation between the mean twin-spacing values determined by TEM and those from the x-ray line profiles for Ag and SiC samples

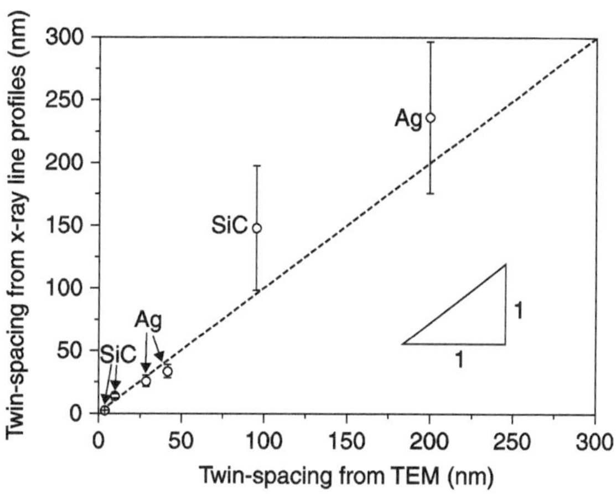

It is noted that the nature of line broadening caused by crystallite size, dislocations and planar faults are essentially different, which enables the separation of their contributions and therefore the determination of the parameters characterising the different lattice defects. This is illustrated in Figure A5, where the full width at half maximum (FWHM) of the diffraction peaks is plotted as a function of the magnitude of the diffraction vector, *g* (Williamson-Hall plot) for the cases where the broadening is caused only by crystallite size (Figure A5a), both size and dislocations (Figure A5b) and both size and twin boundaries (Figure A5c). Figure A5a shows that for defect-free spherical crystallites FWHM is independent of the indices of reflections. When dislocations also contribute to broadening, the breadth of profiles depends on the indices of reflections and the harmonic reflection pairs have different FWHM values (see Figure A5b). Twin boundaries yield the same breadths for harmonic reflections, although the widths of the peaks reflected from planes with various crystallographic orientations (e.g. for 111 and 200 reflections) are different, as illustrated in Figure A5c.

Reliable characterisation of the defect structure can be achieved by fitting the whole x-ray diffraction profiles by theoretical functions calculated on the basis of the model of microstructure. Such powder pattern fitting methods have been elaborated recently [9, 19, 20]. The results presented in this book are obtained mostly by the convolutional multiple whole profile (CMWP) fitting method [20]. In this procedure, the diffraction pattern is fitted by the sum of a background spline and the convolution of the instrumental pattern and the theoretical line profiles related to the crystallite size, dislocations and stacking or twin faults as described previously. In the calculation of the theoretical size profile, the crystallites are modelled by spheres with log-normal size

Figure A5 Full width at half maximum (FWHM) of the diffraction peaks as a function of the magnitude of the diffraction vector, g, for the cases where the broadening is caused by (a) crystallite size alone in Cu, (b) both size and dislocations in Cu and (c) both size and twin boundaries in SiC. The datum points corresponding to harmonic reflection pairs are connected by dotted lines

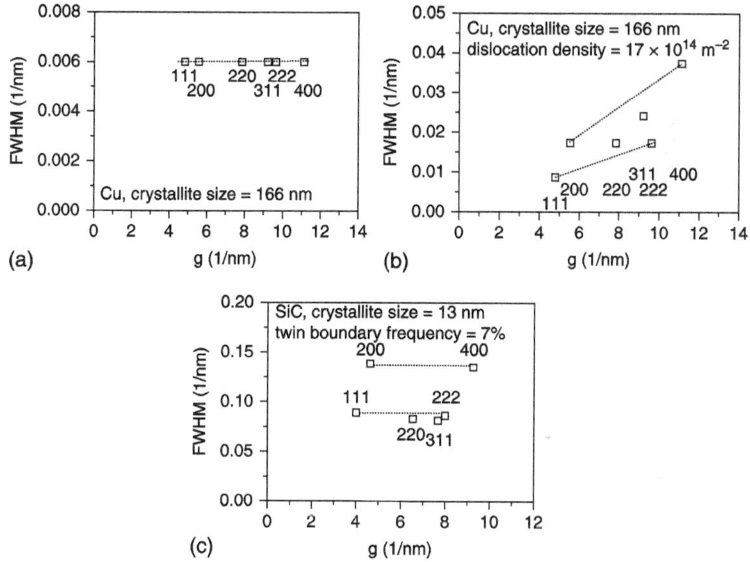

distribution. Figure A6 shows a diffraction pattern for a severely deformed Au sample fitted by the CMWP method. The details of the CMWP procedure are available in Balogh et al. (2006) [17] and Ribárik et al. (2004) [20]. This method gives the following parameters of the microstructure:

1. m, the median of the crystallite size distribution;
2. σ, the variance of the crystallite size distribution;
3. ρ, the dislocation density;
4. R_e, the outer cut-off radius of dislocations. As has been mentioned, instead of R_e, often the dimensionless

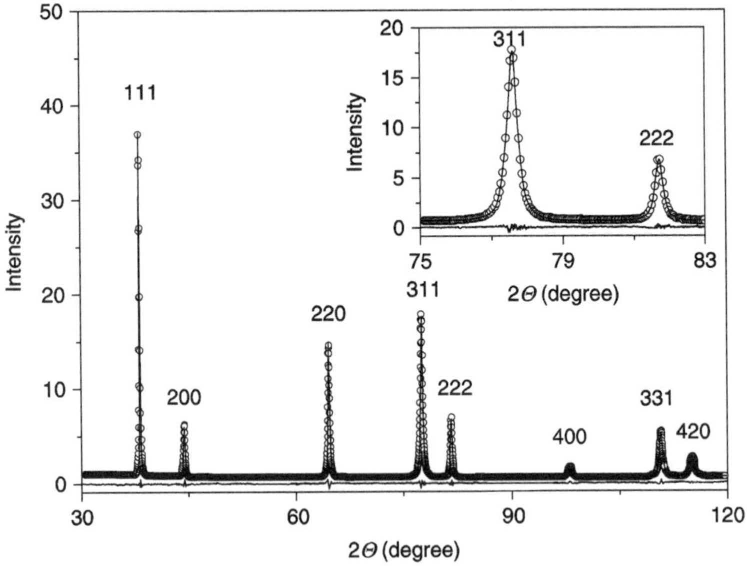

Figure A6 The evaluation of the x-ray diffraction pattern by CMWP method for Au processed by severe plastic deformation. The open circles and the solid line represent the measured and the fitted x-ray diffraction patterns, respectively. A magnified part of the pattern is presented in the inset. The difference between the measured and the fitted patterns is also shown at the bottom of the figure

parameter M is used for the characterisation of arrangement of dislocations;

5. q, or q_1 and q_2 parameters of the dislocation contrast factors for cubic or hexagonal crystals, respectively. These parameters can be used for the determination of the prevailing dislocation slip systems.

From the two parameters, m and σ, of the log-normal size distribution function, $f(x)$, three different mean crystallite sizes can be determined. The arithmetic average of the crystallite diameters ($\langle x \rangle_{arit}$) is obtained as:

$$\langle x\rangle_{arit} = \int_0^\infty x \cdot f(x)dx = m \cdot \exp(0.5\sigma^2), \tag{A15}$$

where x denotes the diameter of crystallites. The mean size of crystallites weighted by their surface areas (referred to as area-weighted mean crystallite size, $\langle x\rangle_{area}$) is given by the following formula:

$$\langle x\rangle_{area} = \frac{\int_0^\infty x \cdot x^2\pi \cdot f(x)dx}{\int_0^\infty x^2\pi \cdot f(x)dx} = m \cdot \exp(2.5\sigma^2). \tag{A16}$$

The mean size of crystallites weighted by their volumes (referred to as volume-weighted mean crystallite size, $\langle x\rangle_{vol}$) can be obtained as:

$$\langle x\rangle_{vol} = \frac{\int_0^\infty x \cdot \frac{x^3\pi}{6} \cdot f(x)dx}{\int_0^\infty \frac{x^3\pi}{6} \cdot f(x)dx} = m \cdot \exp(3.5\sigma^2). \tag{A17}$$

As an example, Figure A7 shows a log-normal crystallite size distribution function for m = 14 nm and σ = 0.45 where the three different mean crystallite size values are also indicated. It is noted that in this book the crystallite size is usually characterised by $\langle x\rangle_{area}$.

X-ray line profile analysis enables the determination of the prevailing dislocation slip systems in hexagonal crystals. The q_1 and q_2 parameters of the dislocation contrast factors obtained by this method depend on the type of dislocations [9]. The 11 possible dislocation slip systems are depicted in Figure A8. The notations of the slip systems used in Figure A8 are listed in Table A2. The 11 dislocation slip systems can be classified into three groups based on their Burgers vectors: $b_1 = \frac{1}{3}\langle\bar{2}110\rangle$ (<a> type), b_2 = $\langle 0001\rangle$ (<c>

Figure A7 **Log-normal size distribution density function, *f*(x), and the arithmetic ($\langle x\rangle_{arit}$), the area ($\langle x\rangle_{area}$), and the volume-weighted ($\langle x\rangle_{vol}$) mean crystallite sizes obtained from *m* and σ**

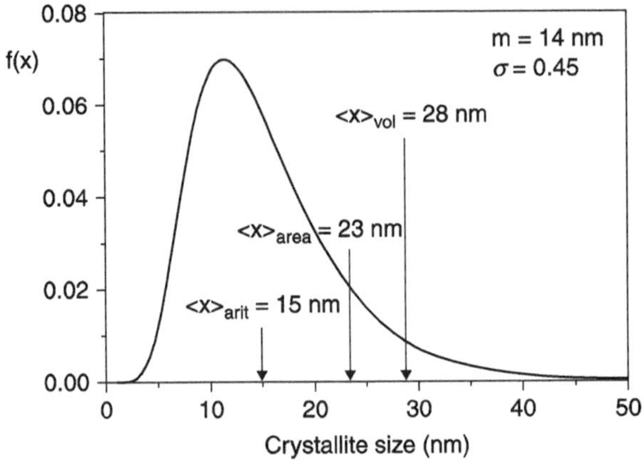

Figure A8 **The 11 possible dislocation slip systems in materials with hexagonal crystal structure. The arrows indicate the three different Burgers vector types: <*a*>, <*c*> and <*c*+*a*>. The slip planes are shown in grey. The Burgers vectors and the slip planes are listed in Table A2**

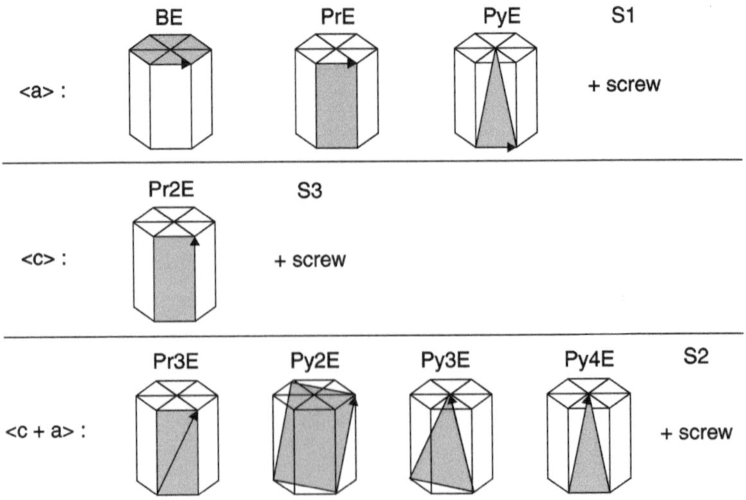

Table A2 The notations, the Burgers vectors and the slip planes of the hexagonal slip systems [16]

Notation	Slip plane and edge/screw character	Burgers vector	Slip plane	Burgers vector type
BE	Basal edge	$\frac{1}{3}\langle\bar{2}110\rangle$	$\{0001\}$	<a>
PrE	Prismatic edge	$\frac{1}{3}\langle\bar{2}110\rangle$	$\{01\bar{1}0\}$	<a>
PyE	Pyramidal edge	$\frac{1}{3}\langle\bar{2}110\rangle$	$\{01\bar{1}0\}$	<a>
S1	Screw	$\frac{1}{3}\langle\bar{2}110\rangle$	–	<a>
Pr2E	Prismatic edge	$\langle 0001\rangle$	$\{01\bar{1}0\}$	<c>
S3	Screw	$\langle 0001\rangle$	–	<c>
Pr3E	Prismatic edge	$\frac{1}{3}\langle\bar{2}113\rangle$	$\{01\bar{1}0\}$	<c+a>
Py2E	Pyramidal edge	$\frac{1}{3}\langle\bar{2}113\rangle$	$\{2\bar{1}\bar{1}2\}$	<c+a>
Py3E	Pyramidal edge	$\frac{1}{3}\langle\bar{2}113\rangle$	$\{11\bar{2}1\}$	<c+a>
Py4E	Pyramidal edge	$\frac{1}{3}\langle\bar{2}113\rangle$	$\{01\bar{1}1\}$	<c+a>
S2	Screw	$\frac{1}{3}\langle\bar{2}113\rangle$	–	<c+a>

type) and $b_3 = \frac{1}{3}\langle\bar{2}113\rangle$ (<*c* + *a*> type). There are 4, 2 and 5 slip systems in the <*a*>, <*c*> and <*c*+*a*> Burgers vector groups, respectively. The theoretical q_1 and q_2 values for the 11 possible slip systems in hexagonal crystals can be calculated according to the formulas presented in Borbély et al. (2003) [15] and Kuzel and Klimanek (1989) [21] and by using the software *Anizc* on the website *http://metal.elte.hu/anizc*. The theoretical values of q_1 and q_2 for some hexagonal materials are listed in Dragomir and Ungár (2002) [16]. Comparing the measured values of $q_1^{(m)}$ and $q_2^{(m)}$ with the theoretical values calculated for the 11 slip systems, the prevailing slip systems in the studied material can be obtained. A computer program was elaborated to facilitate the determination of the Burgers vector population from the dislocation contrast factors [22]. First, the software selects some slip systems from one of the three Burgers vector groups. In the second step, the theoretical $\bar{C}_{hk0}q_1$ and $\bar{C}_{hk0}q_2$ values of the selected slip systems are averaged with equal weights. This procedure is carried out for each Burgers vector group. The relative fractions of the three groups, h_i ($i=1,2,3$) are calculated by solving the following three equations:

$$q_1^{(m)} = \frac{1}{P}\sum_{i=1}^{3} h_i b_i^2 \left(\overline{\bar{C}_{hk0}q_1}\right)_i, \tag{A18}$$

$$q_2^{(m)} = \frac{1}{P}\sum_{i=1}^{3} h_i b_i^2 \left(\overline{\bar{C}_{hk0}q_2}\right)_i, \tag{A19}$$

and

$$\sum_{i=1}^{3} h_i = 1 \tag{A20}$$

where $\left(\overline{\bar{C}_{hk0}q_1}\right)_i$ and $\left(\overline{\bar{C}_{hk0}q_2}\right)_i$ are obtained by averaging for the *i*th Burgers vector group and P is given as $P = \sum_{i=1}^{3} h_i b_i^2 (\overline{\bar{C}_{hk0}})_i$.

If the three h_i weights have positive values, the program stores them as one of the possible solutions. After examining all the possible combinations of the slip systems, ranges of the three weights are obtained as the final solution. The detailed description of the procedure is given in Máthis et al. (2004) [22].

References

1. A. Leineweber and E.J. Mittemeijer, 'Anistropic microstain broadening due to compositional inhomogeneities and its parametrisation', *Zeitschrift für Kristallographie*, Suppl. 23 (2006) 117–122.
2. M. Leoni and P. Surface, 'Scardi Relaxation Effects in Nanocrystalline Powders', In: E.J. Mittemeijer and P. Scardi (eds), *Diffraction Analysis of the Microstructure of Materials*, Springer Verlag, Berlin (2004), pp.413–454.
3. T. Ungár, E. Schafler and J. Gubicza, 'Microstructure of Bulk Nanomaterials Determined by X-Ray Line Profile Analysis', In: M.J. Zehetbauer and Y.T. Zhu (eds) *Bulk Nanostructured Materials*, Wiley–VCH, Weinheim (2009) pp.361–386.
4. J. Gubicza and T. Ungár, 'Nanocrystalline materials by powder diffraction line profile analysis', *Zeitschrift fur Kristallographie*, 222 (2007) 567–579.
5. E.F. Bertaut, 'Raies de Debye-Scherrer et repartition des dimensions des domains de Bragg dans les poudres polycrystallines', *Acta Crystallographica*, 3 (1950) 14–18.
6. A. Guinier, *X-ray Diffraction*, Freeman, San Francisco, CA. (1963).
7. B.E. Warren, *X-ray Diffraction*, Dover Publications, New York (1990).
8. G. Ribárik, T. Ungár and J. Gubicza, 'MWP-fit: a program for multiple whole-profile fitting of diffraction

peak profiles by ab initio theoretical functions', *Journal of Applied Crystallography*, 34 (2001) 669–676.
9. T. Ungár, J. Gubicza, G. Ribárik and A. Borbély, 'Crystallite size distribution and dislocation structure determined by diffraction profile analysis: principles and practical application to cubic and hexagonal crystals', *Journal of Applied Crystallography*, 34 (2001) 298–310.
10. M.A. Krivoglaz and K.P. Ryaboshapka, 'The effect of dislocations on line profiles', *Fizika Metallov i Metallovedenie*, 15 (1963) 18–27.
11. M. Wilkens, 'Fundamental Aspects of Dislocation Theory', Ed.: J.A. Simmons, R. de Wit, R. Bullough, Vol. II. Nat. Bur. Stand. (US) Spec. Publ. No. 317, Washington, DC. USA, (1970) pp.1195–1221.
12. I. Groma, T. Ungár and M. Wilkens, 'Asymmetric X-ray line broadcasting of plastically deformed crystals. I. Theory', *Journal of Applied Crystallography*, 21 (1988) 47–54.
13. T. Ungár and A. Borbély, 'The effect of dislocation contrast on x-ray line broadening: A new approach to line profile analysis', *Applied Physics Letters*, 69 (1996) 3173–3175.
14. T. Ungár and G. Tichy, 'The Effect of Dislocation on X-Ray Line Profiles in Untextured Polycrystals', *Solid State Physics*, (a) 171 (1999) 425–434.
15. A. Borbély, I. Dragomir-Cernatescu, G. Ribárik and T. Ungár, 'Computer program ANIZC for the calculation of diffration contrast factors of dislocations in elastically anistropic cubic, hexagonal and trigonal crystals', *Journal of Applied Crystallography*, 36 (2003) 160–162.
16. I.C. Dragomir, T. Ungár, 'Contrast factors of dislocations in the hexagonal crystal system', *Journal of Applied Crystallography*, 35 (2002) 556–564.

17. L. Balogh, G. Ribárik and T. Ungár, 'Stacking faults and twin boundaries in fcc crystals determined by x-ray diffraction profile analysis', *Journal of Applied Physics*, 100 023512 (2006).
18. L. Balogh, G. Tichy and T. Ungár, 'Twinning of pyramidal planes in hexagonal close packed crystals determined along with other defects by X-ray line profile analysis', *Journal of Applied Crystallography*, 42 (2009) 580–591.
19. P. Scardi and M. Leoni, 'Foundations of Crystallography', *Acta Crystallographica*, A58 (2002) 190–200.
20. G. Ribárik, J. Gubicza and T. Ungár, 'Correlation between strength and microstructure of ball-milled AI-Mg alloys determined by X-ray diffraction', *Materials and Science Engineering*, A387–389 (2004) 343–347.
21. R. Kuzel and P. Klimanek, 'X-ray diffraction line broadening due to dislocations in non-cubic crystalline materials. III. Experimental results for plastically deformed zirconium', *Journal of Applied Crystallography*, 22 (1989) 299–307.
22. K. Máthis, K. Nyilas, A. Axt, I.C. Dragomir, T. Ungár and P. Lukác, 'The evolution of non-basal dislocations as a function of deformation temperature in pure magnesium determined by X-ray diffraction', *Acta Materialia*, 52 (2004) 2889–2894.

Index

FSC
www.fsc.org
MIX
Papier | Fördert gute Waldnutzung
FSC® C083411

Kolibri 360 GmbH
Ferdinand-Jühlke-Straße 7
99095 Erfurt, Deutschland
produktsicherheit@kolibri360.de